C
TRAVELS

CLIMATE TRAVELS

HOW ECOTOURISM CHANGES MINDSETS AND MOTIVATES ACTION

MICHAEL M. GUNTER, JR.

Columbia University Press *New York*

Columbia University Press
Publishers Since 1893
New York Chichester, West Sussex
cup.columbia.edu

Library of Congress Cataloging-in-Publication Data
Names: Gunter, Michael M., 1969– author.
Title: Climate travels : how ecotourism changes mindsets and motivates action / Michael M. Gunter, Jr.
Description: New York : Columbia University Press, [2023] | Includes bibliographical references and index.
Identifiers: LCCN 2022032557 (print) | LCCN 2022032558 (ebook) | ISBN 9780231205887 (hardback) | ISBN 9780231205894 (trade paperback) | ISBN 9780231556217 (ebook)
Subjects: LCSH: Climate change mitigation—United States—Citizen participation | Ecotourism—Social aspects—United States | Travel—Environmental aspects
Classification: LCC TD171.75 .G86 2023 (print) | LCC TD171.75 (ebook) | DDC 363.738/74—dc23/eng/20221019
LC record available at https://lccn.loc.gov/2022032557
LC ebook record available at https://lccn.loc.gov/2022032558

Columbia University Press books are printed on permanent and durable acid-free paper.
Printed and bound by CPI Group (UK) Ltd, Croydon, CR0 4YY

Cover design: Noah Arlow
Cover image: Shutterstock (Mt. Rainier)

For Linda

CONTENTS

II DO IT YOURSELF: ACTION MAKING A DIFFERENCE

ACKNOWLEDGMENTS

This work is a tribute to over 125 interviewees who agreed to tell their climate change stories, both good and bad. While there are far too many to name individually, I thank these climate scientists, biologists, foresters, eco-entrepreneurs, business executives, and architects for their patience and dedication. Similarly, city managers and mayors, public works directors and civil engineers, environmental activists, journalists, political scientists, economists, and historians all made this work stronger. Graciously sharing their personal perspectives, these professionals help explain the complex climate narrative we must all face.

Special thanks to my agent Jane Dystel at Dystel, Goderich & Bourret for shepherding this project in its early stages and finding the perfect home for it at Columbia University Press. This book is immeasurably better thanks to her professional expertise, along with her colleagues Miriam Goderich and Melissa Melo. At Columbia University Press, Miranda Martin and Brian Smith were early champions for this work, while Leslie Kriesel, Noah Arlow, and Gregory McNamee were indispensable during production stages. I also wish to thank Robyn

Massey and Meredith Howard for their expertise and passion as well as indexer Fred Leise for his keen eye.

At Rollins College, Provost Susan Singer and Dean Jennifer Cavenaugh generously supported multiple stages of this work. I deeply thank my administrative assistants, Alison Reeve and Nakia Gater, who kept me on track with other teaching and administrative responsibilities throughout this research and writing. Karla Knight and Tina Hall deserve special mention as well. Funding for this project came from a McKean Grant and multiple Critchfield awards here at Rollins as well as an Elizabeth Morse Genius Foundation grant secured by my colleague Bruce Stephenson. And it was a Fulbright Scholar award, years ago in the Slovak Republic, that first opened my eyes to the power of travel.

Words cannot express the gratitude I owe to my father, Michael Gunter, and my late mother, Judy Gunter, for their advice, love, and support. Dan and Nancy Kula also warmly shared their time and love. Personal and professional friends Tom and Sally Lairson hold a special place in the heart of our family, as does my good friend Richard Cooper. Finally, this book is a tribute to the love of Linda and our children, Ansleigh, Malachi, and Emerson. You are the inspiration for the words that follow.

INTRODUCTION

This Land Is Your Land

Our country is in trouble. From its redwood forests to its Gulf Stream waters, as Woody Guthrie once poetically wrote, this land, like the rest of our planet, faces a climate crisis. It's true that climate has changed in the past, that global warming and cooling periods repeated roughly every 100,000 years for at least the last millennium. But today is different. We should be in a cooling phase right now, yet instead, the earth is warming. And it's warming an astounding ten times faster than at any period in the last 66 million years.

Anthropogenic greenhouse gases are the primary culprit. Since the Industrial Revolution, we've pumped more than 500 billion tons of heat-trapping carbon dioxide into our atmosphere.[1] In only a little more than three decades, since 1988, we've emitted more industrial carbon than in all prior human history.[2] In my lifetime alone, we've become much more than a biological force. We're a geological force. Our emissions have changed the very chemistry of our atmosphere to create a new epoch, what Nobel Prize–winning chemist Paul J. Crutzen and his marine science collaborator Eugene F. Stoermer label the Anthropocene.

Scholars debate whether this is a moral or economic problem, a technological or behavioral problem, a national security or an international security problem. Truth be told, it's all of those wrapped together. Climate change is complex and nonlinear, subject to rapid, unpredictable shifts. Its causes and its impacts intertwine as soot from California wildfires, themselves often with connections to climate change, increases melting of glaciers in Greenland by darkening its ice and enhancing the absorption of heat. The result is, to quote esteemed environmental activist and author Bill McKibben, who penned the first popular book on climate change in 1989, a "crescendo of cascading consequences."[3]

Scientists warned an American president of these dangers back in 1965, and the public was alerted well over a quarter-century ago. Yet to date, national-level action within the United States can be characterized only as woefully inadequate, as emissions are 62 percent higher than when international climate negotiations began in 1990.[4] "We're suffering through the consequences of decades of inaction," Michael Brune lamented to me one summer afternoon. As the former executive director of the Sierra Club, our nation's oldest and largest environmental organization, Brune draws upon a wealth of experience and expertise, adding, "We will suffer from climate change that is already here and will get worse no matter what."[5]

National leadership has failed us, plain and simple. That is unfortunate—but not surprising. As Michael Bloomberg, former New York City mayor and United Nations Special Envoy for Cities and Climate Change, contends, Congress is not well positioned to lead on such matters. It is an institution designed to follow, reacting only when external forces like the 1973 Arab oil embargo inflict sufficient pain. On revolutionary issues, from women's suffrage to civil rights to gay rights, voters consistently lead politicians instead of the other way around.

That means the ball is in our court. If we want more from our elected officials on climate change, we must let them know. In our democratic republic, government reflects public will. With an eerily similar focus on environmental security, former President Jimmy Carter experienced a twist on this frustration firsthand following his July 1979 crisis-of-confidence speech. "For your good and for your nation's security," President Carter urged Americans to vow "to take no unnecessary trips, to use carpools or public transportation whenever you can, to park your car one extra day per week, to obey the speed limit, and to set your thermostats to save fuel. Every act of energy conservation like this is more than just common sense—I tell you it is an act of patriotism."[6] Many later faulted this address, ridiculed as the "malaise" speech, and in part, blamed it for Carter's failed 1980 reelection bid against Ronald Reagan.[7] Americans did not see his appeal as patriotic. They did not view such actions as in our collective, let alone individual, interest.

To better address our climate crisis, that mindset must change. We must understand that addressing climate change is in both our individual and collective interests. But for that to occur, we must become more knowledgeable about what climate change really is. We must better understand how it threatens us here in the United States. And we must restructure the energy choices and daily living habits that drive those threats. Moving deftly through formidable amounts of political and scientific literature, along with well over one hundred interviews of climate scientists and activists, public servants and entrepreneurs, this book offers a start in that critical educational process.

As a complex, multifaceted problem, climate change has no single solution. Author and activist Paul Hawken spells out 100 preferred options with his Project Drawdown team, born out of a Seattle town hall where he casually asked two teenagers next to him who the top ten NBA teams were. After those kids quickly rattled off a list, Hawken started thinking, "No one can do that with climate solutions—and that's a problem."[8] He then assembled a research team to identify and measure 100 solutions to carbon drawdown over the next three decades. A 120-person advisory board of geologists, engineers, agronomists, politicians, writers, climatologists, biologists, botanists, economists, financial analysts, architects, and activists regularly review this living plan, which is updated online. Alas, while many of the initiatives Hawken and his team identify are under way, the political will to scale them up remains insufficient. The pages that follow home in on this deficiency, outlining the most promising political roadmap to address climate change more effectively.

This comes in two parts.

First, Americans must see what climate change along our home front looks like. That's part I of this book, describing eight core climate threats unfolding across this land. Hearing about climate change from scientists and the media that report their findings has not been enough to motivate most Americans to action. We need to see the threats ourselves. This doesn't require traveling to the far corners of our country or witnessing all eight threats outlined in this book. Travel may be just around the corner from your neighborhood, where the threat most persistent in your region resides.

But bearing witness is not enough. After seeing climate change in action, Americans must act. That's the second half of this book. In the spirit of activists like Paul Hawken, but emphasizing the local level, these pages stress seven categories of action already under way, categories you as an individual can help scale up within your own community. Individual, local acts will not be enough, of course, when it comes to solving global climate change. Yet, national and international efforts, to date, fall woefully short precisely because they lack sufficient local support.

Direct experience with both threats and solutions, as this book outlines, can change that equation.

Our best option to tackling the climate crisis, then, lies at the local level, aided by travel experiences that broaden understanding of our surroundings—and how our daily actions affect them. This builds on age-old arguments from philosophers Jean-Jacques Rousseau and John Dewey to travel writers Mark Twain and Rick Steves: direct experience is a powerful educational tool. It helps us better understand not only places farther afield but also the homes from which we come. It brings our own strengths into sharper focus. And it inspires creative ideas to address our weaknesses.

Of course, if not done conscientiously, travel exacerbates our carbon problem. Where one travels matters. How one travels does, too. And perhaps most important, action taken afterward is the ultimate test as to whether benefits outweigh the harms from travel. Along these lines, if travel inspires some form of grassroots climate activism, it also offers salvation for our struggling democracy by expanding democratic participation and promoting political innovation. That activity in and of itself confronts another debilitating obstacle when it comes to climate change: eco-anxiety. Defined by the American Psychological Association as "the chronic fear of environmental doom," this all-too-common emotional response to consistently worrisome environmental news often short-circuits individual initiative. Those who truly understand the climate crisis feel too overwhelmed to act.

But do not despair. While selfless philanthropists and superhero politicians may assist, all we really need is you. To effectively engage, though, you must better understand why a culture of climate apathy persists in many corners of our country. Lack of information is partly responsible, misinformation even more so. Scholars identify this phenomenon as the information deficit model or science comprehension thesis and point to public limitations concerning scientific understanding. Noted authors like Naomi Oreskes and Erik M. Conway detail a concerted campaign to obscure the truth on climate change, with financial as well as ideological interests driving rationale.[9] This influence cannot be understated, from Exxon to the Heartland Institute to the Koch brothers. "Machiavelli warned about being an agent of change, that very few people survive that," cautions Phillip Stoddard, a biology professor at Florida International University and former mayor of South Miami, who experienced those pressures firsthand during his multiple-term tenure. "People invested in the status quo will fight you every inch of the way."[10]

But that is not the only hurdle climate activists face. At the risk of undermining rationality altogether, climate communications research suggests additional, less heady factors are even more crucial. Tapping into emotions with personal stories, these researchers assert, proves particularly effective in countering climate apathy.[11] That approach echoes the perspective of renowned environmental author Aldo Leopold, who believed that direct contact with the natural world was crucial to extending personal ethics beyond our own self-interest. "When we see land as a community to which we belong, we may begin to use it with love and respect," he wrote.[12]

Think about it. We simply cannot love what we don't really know. "When people are close to the land, they can make decisions that are informed. They understand the consequences," echoes Michael W. Klemens, a conservation biologist with the American Museum of Natural History who also pursues interests in land-use and public policy. "In a technological society, we can get so divorced from nature and don't understand the consequences. When people experience it firsthand, they can advocate for it."[13] Stoddard agrees. "The basic rule is don't come in on a white horse to solve a problem that people don't know is a problem," he says. "You have to sell the problem before you can sell the solution, even if you have been living and breathing that problem for years. If everyone else doesn't see the problem, you are not going to solve it."[14]

This is the missing link in the struggle against climate change today.

Scientific reports filtered through traditional media sources and even modern social media have not changed the minds of a significant subset of Americans. Even worse, these reports harden ideological resistance, despite continued advances in scientific understanding over the past two decades.[15] Conscientious travel, with a minimal ecological footprint, offers a different approach. It taps into innate human curiosity about our planet. It breaks down barriers and widens perspectives. Travelers see for themselves the negative effects of climate change—and the positive actions making a difference. This strengthens civil society, which will not only spur effective climate change mitigation but also decrease the political polarization threatening our democracy today.

Best-selling author Rachel Carson, widely recognized as the founder of modern environmentalism, tapped into the power of emotions much like Leopold to offer further insight. She emphasized a "sense of wonder" as central to understanding our world. We all had that sense of wonder as children, but too often we take our world for granted as adults. To give us a chance with climate change, we must heed the advice of Carson and Leopold as well as Klemens and Stoddard. We must better appreciate our surroundings. And like any loving relationship, that means we cannot take Earth for granted. Absence may make the heart grow fonder, but you must return to your loved one periodically for that fire to persist. We must see this land to love it.

We don't do that enough today. We haven't really done a good job of returning to our natural roots for a long time, particularly during the Great Acceleration of carbon dioxide emissions over the last half-century. Even our children suffer from what author and columnist Richard Louv calls nature deficit disorder.[16] Borrowing from his discussion of children's struggles with depression, obesity, and attention deficit disorder, the prescription regarding climate apathy is similarly straightforward: more time outside. Nature outings need not be on the level of American naturalist and philosopher Henry David Thoreau to foster introspection. A simple family summer vacation to Glacier National Park, still home, at least at this writing, to twenty-five of its previously 150 glaciers in the mid-nineteenth century, will suffice. The important point is you keep that spark alive, that sense of wonder Rachel Carson identified. Only then, as Oberlin College environmental professor David Orr argues, can we rebuild our civic intelligence, extending notions of citizenship to include the environment as well.[17]

It is true that technological advances will drive the energy revolution, with shifts to fuels from heaven instead of those from hell, as Rochelle Lefkowitx, president of Pro-Media Communications, observes.[18] Wind and solar energy will increasingly replace fossil fuels. But it is dangerous to think of our climate crisis as simply a technological problem. "Technology neither requires nor results in any particular improvement in behavior, politics, or economics that brought us to our present situation in the first place," Orr argues.[19] As recognized in award-winning journalist Naomi Klein's acclaimed critique of capitalism, our climate crisis is not rooted merely in economics, either. Our problem is economic and technical. But it's also cultural and political. The very terms *climate change* and *global warming* can kill conversation within surprisingly too many circles, even decades after international recognition of our high stakes. Florida's former governor and now U.S. Senator Rick Scott, for example, forbade state officials from using those words during his gubernatorial tenure.[20] How are we to build a bipartisan political base addressing climate change when we can't even utter that phrase?[21] And more immediately, what can you, just one individual, do to make a difference?

Former U.S. Congressman Bob Inglis, a Republican, has a suggestion. It centers on not facts themselves but on where those facts come from. "This is not a head problem," asserts Inglis. "This is a heart problem."[22] Inglis, who received the 2015 John F. Kennedy Profile in Courage Award, is executive director of republicEn.org, a conservative organization dedicated to addressing climate change, and is an advisory board member to the nonpartisan Citizens' Climate Lobby. He believes a conservative answer to climate change exists. "We are part of what we call the eco-right," Inglis explains. "What I like to tell conservatives is we failed to present an alternative to social security, Medicare, Medicaid, Obamacare. Each time our country took something over nothing. But conservatism has muscular solutions to real-world problems. This is an opportunity to prove the power of private enterprise."[23]

Inglis did not always take that position. He initially opposed efforts to address climate change. As he describes, Inglis represented the reddest district in the reddest state in the nation, the Fourth Congressional District of South Carolina. After a string of terms from 1993 to 1999, followed by a six-year stint practicing commercial real estate law, he won the same House seat to serve again from 2005 to 2011. But he took a much different position on climate change that second time around. Credit goes first to his kids. They told him, "Dad, I'll vote for you, but you have to clean up your act on the environment." Inglis listened. And then he traveled, learning about the impact of climate change firsthand in his position as the ranking member of the House Science Committee's Energy and Environment Subcommittee. Those direct experiences, with a push from his five children, persuaded him to change his climate position.

Again, all travel comes with greenhouse gas emissions, but many of those carbon costs shrink when green techniques are applied. It matters how you travel, how you behave at your destination, and, most important, how you adjust your daily life afterward. As Bill McKibben asserts, we must make our travel count. This combination of witnessing impacts of climate change yourself and prioritizing local actions serves as the best roadmap to solving our climate crisis. That blend is what best builds both political salience and efficacy. With that in mind, let's look more closely at the climate threats to our home front.

I

SEE IT YOURSELF

Threats to the Home Front

1

OUR RISING SEAS

There's fish swimming in this street. As I stroll along Surrey Crescent, a block off Norfolk's main drag of Hampton Boulevard, evidence of sea level rise surrounds me. Skip Stiles, executive director of the statewide conservation group Wetlands Watch, invited me here to see several climate adaptations this Virginia city is employing. We're within the affluent Larchmont neighborhood, the linchpin of the municipal tax base, and surrounding waterfront homes offer stunning views. Many of them are also increasingly vulnerable to sea level rise. Norfolk's Coastal Resilience Strategy recognizes this, pursuing projects from Ohio Creek to the Lafayette River to Mason Creek. Stiles is showing me an early example off a finger of the Lafayette River. Broad green spaces interspersed with majestic oaks were Myrtle Park's signature—until the sea started moving in.

Repeated tidal flooding cracked and buckled portions of the low sea wall on its edge. Ponding saltwater killed vegetation, leaving mud and instability in its wake. After a welcome wetlands makeover in 2014, marsh grasses now line nearby streets. An arched walking bridge traverses still more of these grasses as well as a shifting tidal pool. Neighborhood children and dogwalkers frequent Myrtle Park once again. Yet fish still swim in this connecting street. The Myrtle Park Wetland Restoration Project bought this neighborhood time, but the sea is still rising. As city landscape equipment blares in the background on this hot and sunny August afternoon, Stiles offers more details. "We spent $300,000 to restore this marsh, plus another $1.2 million to raise the street eighteen inches," explains Stiles. "But that's not going to be enough. This area has the highest sea level rise on the East Coast."[1]

Global sea levels have risen about nine inches since we started keeping track in 1880. By 2100, additional sea level rise could displace 13 million Americans.[2] Climate change is a big reason why. But subsidence, the sinking of land, factors in as well. In fact, along our eastern coast, subsidence combines with ocean current changes tied to climate change to push sea level rise 50 percent higher than the global average. Measurements since 1927 from the National Oceanic and Atmospheric Administration tidal gauge at nearby Sewell's Point show the highest relative sea level rise along our entire eastern coast, nearly fifteen inches.

Not surprisingly, all four main impacts from sea level rise challenge Norfolk and the larger Hampton Roads region: shoreline erosion, storm intrusion, salinization, and flooding. These impacts will continue for centuries, even if we stabilize global mean temperature tomorrow. There's just too much carbon baked into the system. As the Intergovernmental Panel on Climate Change (IPCC) ominously asserts, it is "virtually certain that global mean sea level rise will continue for many centuries beyond 2100, with the amount of rise dependent on future emissions."[3]

Stiles began working on sea level issues in 2007, focusing expressly on the local level. "Because local government controls all the factors that need to adapt, land use, business permits, even administering most federal programs," he explains. "Wetlands Watch brings science to local governments in a way that fits their decision making. We work hand in glove with local governments." Stiles's point is well taken. Local threats like this, driven by global climate change, resonate more effectively than fears about international impacts. They also encourage a wider political tent to consider policy solutions. Instead of shutting down conversations, local threats foster them, highlighting areas those in the local community treasure. "We don't push climate change," he continues. "We use it to get into the conversation, reframed as saving communities from flooding. Now it's not just the coastal stuff. It's everywhere. We're making much more progress . . . It's turned from an ideological issue to a constituent issue. I'm wet, and I'm pissed off."[4]

Across the globe, coastal cities like Norfolk, totaling two thirds of worldwide population, are vulnerable. A recent study estimates 340 million people are at risk of annual flooding in the next thirty years.[5] China, India, Bangladesh, Vietnam, Indonesia, Japan, Egypt, Thailand, and the Philippines are among the ten countries with the most people living in vulnerable coastal zones. The United States is another. It's not just cities at risk. Farmlands, fisheries, and rice paddies are as well. Louisiana lost nearly 1,900 square miles of land between 1932 and 2000, according to the U.S. Geological Survey. That's roughly the size of Delaware. And the Pelican State continues to lose land faster than almost anywhere, a football field's worth each hour.[6] You won't even need to be on the coast to face risks in the decades to come. With a five foot sea level rise, our National Mall in Washington, 120 miles from the ocean, will flood.

Sea levels are not simply rising. They are rising at faster and faster rates. Sophisticated satellite technology allows scientists to identify annual rise within tenths of a millimeter. Rates since the mid-nineteenth century are larger than the mean rate from our previous two millennia. Before 1990, according to NASA, the rate was 1.1 millimeters per year. From 1993 to 2012 it was 3.1 mm per year. Since then, it has been 3.4 mm.[7]

Several factors shape this. One is the aforementioned subsidence. When we remove fluids from the ground, from groundwater to oil, we also remove what props that land up. Much of our eastern seaboard faces this, from Miami Beach to Norfolk to New York and Boston.[8] Another factor, one in our favor, is something called glacial isostatic rebound. This refers to land rising as the weight of ice melts away. Waters in Puget Sound, for example, may rise another foot or more this century. But the Olympic coast will probably see a net sea level rise of only about two inches, according to a 2008 report by the University of Washington Climate Impacts Group and Washington Department of Ecology, thanks to glacial isostatic rebound.

Beyond that, at least four other factors relate to climate change.

Gravity is one explanatory variable. Greenland's ice sheet is so large it has its own gravitational pull, attracting water toward it. As this ice melts, corresponding weight loss means its gravitational pull weakens and ocean waters will shift accordingly. This regional effect is called fingerprinting—and it's bad news for American cities like New Orleans, Miami, Charleston, Washington, Baltimore, New York, Boston, and Norfolk.

Another factor is a slowing Gulf Stream, which is at its weakest flow in at least 1,600 years. This pushes more water up against our coasts.[9] We know the Gulf Stream, formally referred to as the Atlantic Meridional Overturning Circulation (AMOC), for its warm waters flowing from the tip of Florida northward along the eastern coastline of the United States and Canada. Europeans know it for its moderating influence on continental weather, making Europe's western portion much warmer than it otherwise would be. What drives this conveyor belt of water is the density differential within Atlantic Ocean waters. Salty, warm water along Florida's coast moves north and cools. This cooling makes it denser, and, when it's heavy enough, that water sinks and flows back south. As Greenland melts, adding fresh water and reducing the salinity of the Atlantic, it also makes this water less dense, which slows the traditional sinking and thus the movement of the AMOC.

A third factor is thermal expansion. It's responsible for roughly one-third of global sea-level rise observed by satellite altimeters since 2004, according to NASA.[10] Water expands as it warms, taking up more space than cold water. If you heat fifty gallons of water to 100 degrees Fahrenheit, it grows to roughly fifty-one gallons. With average ocean depth over two miles, even minimal thermal expansion translates into measurable sea level rise. Climate scientist James Hansen predicts that sea levels will rise a foot from this thermal expansion if global temperatures increase 2.7 to 8.1 Fahrenheit (1.5 to 4.5 degrees Celsius).

Fourth, most troublesome of all, we add more water to our oceans as Greenland and Antarctica melt. Just exactly how much we add depends on how quickly that melt happens. And much uncertainty surrounds this question.

Ice ages come and go. We've been alternating between long ice ages and shorter interglacial periods for around 2.6 million years. Over the last million-odd years, the rotation has occurred approximately every 100,000 years, with around 90,000 years of ice age followed by 10,000 years of warm period we call an interglacial. The Holocene interglacial lasted about 11,700 years, suggesting that the next cooling phase should already be under way, along with its corresponding fall in sea level. Instead, as the late atmospheric scientist Paul Crutzen revealed during a 2000 conference, we've changed the expected cycle, moving into something he labeled the Anthropocene to highlight human impacts on our planet.[11]

That said, the obvious question becomes what causes ice ages?

The answer is complicated, including changes in Earth's orbit around the sun, along with changes in ocean circulation and atmospheric composition. Milankovitch Cycles are a major influence, leading us into ice ages every 95,000 to 125,000 years. They are named after Serbian mathematician Milutin Milankovitch and his thirty years of hand calculations from 1911 to 1941.[12] Our last ice age peaked around 20,000 years ago, after which sea level rose roughly four hundred feet over a 14,000-year period before remaining relatively steady for the last 6,000 to 8,000 years. Greenland's ice sheet is a holdover from this last ice age.[13]

Antarctica represents even greater uncertainty. The coldest, windiest, and highest continent, on average, it's also, surprisingly, our driest. There's little snowfall except along its coasts. As a result, most of Antarctica is officially a desert. I had to keep reminding myself that during a weeklong excursion over a decade ago, especially considering an ice sheet covers 98 percent of the continent. That ice averages over one mile in thickness, with some portions more than twice that, meaning Antarctica will likely take hundreds of years to melt. When it all does, this will raise sea level about two hundred feet.

But there's a key difference in East and West Antarctica. The East Antarctic ice sheet is much thicker, nearly three miles in some parts, and largely rests on bedrock above sea level. That means it is less susceptible to warming waters around the continent and relatively stable. The West Antarctic ice sheet, on the other hand, is nowhere near as thick—and much of its underlying bedrock is below sea level. These submerged portions are vulnerable to changes in water temperatures and, thus, considerably more unstable. They will melt first.

If all remaining ice cover melts, seas will rise around 240 feet. Greenland contributes approximately twenty-two feet, West Antarctica another twenty, and East Antarctica nearly two hundred. All alpine glaciers combined may add another foot and a half. This will likely take centuries, with most models suggesting a maximum of six to seven feet sea level rise by 2100.[14] But that's enough to wipe Pacific Island nations like Tuvalu off the map, while rich countries like the United States spend billions to save select coastlines. And it means the remnants of cities like Norfolk will look decidedly different.

This is particularly true if esteemed climate scientists like Hansen are correct. Hansen fears that sea-level rise models topping out at six or seven feet this century are far too conservative, failing to consider exponential melt increase in Antarctic and Greenland. Most climate models assume change unfolds gradually. But geological records reveal rapid pulses do occur.[15] Some 14,000 years ago, for example, seas rose 13 to 16.4 feet per century over several centuries. About 120,000 years ago, they rose 6.5 to 9.8 feet in fifty years or less.[16]

Let's look a little more closely at Norfolk, one of nine cities and seven counties in the Hampton Roads metropolitan area. It's Virginia's second-largest city, home to nearly 250,000 people as well as the world's largest naval base. Surrounded by water on three sides with the Chesapeake Bay and the Elizabeth and Lafayette Rivers, Norfolk has always experienced flooding from heavy rains as well as storm surge from nor'easters and hurricanes. But sea level rise makes matters even worse. Annual flooding events have tripled since the 1970s. Tidal flooding now occurs about once a month—and may more than triple that by the end of the 2020s.

Norfolk already requires that new construction be three feet above the 100-year flood level. Fugro, a Dutch energy infrastructure firm, recently recommended over $1 billion in adaptation measures, including floodwalls, tide gates, elevated roads, and pumping stations.[17] An even bigger headline for Norfolk revolves around Norfolk Naval Station, which covers more than 6,000 acres and employs 82,000 people, providing $50 billion in revenue to the Hampton Roads region. Sea levels will rise here between three to five and a half feet by the end of the next century, according to projections from the Virginia Institute of Marine Science. By 2041, Hampton Boulevard, the main road into the base, just a block from Myrtle Park and the fish swimming in the street, will be impassable for two to three hours daily during high tide, according to Larry Atkinson, an oceanographer at Norfolk's Old Dominion University.[18] Most notably, the centerpiece of the base, fourteen piers that support seventy-five ships, is extremely vulnerable to tidal and storm-induced flooding. Addressing that will cost $60 million per pier.[19] When the U.S. military starts spending sums like that on climate adaptation rather than weapons acquisition, we all should take notice.

Beyond affecting our military readiness, rising sea levels also have profoundly unequal societal impacts. "Is it the direct risk, the water at your front door, or is it the way water reverberates through markets and racism?" asks Katharine Mach, professor of environmental science and policy at the University of Miami. "It's not just a question of who floods, but how this shifts development patterns, speculation and markets, foreign investment. All those things come into consideration."[20] We've already seen that on the Louisiana coast at Isle de Jean Charles, where many of its residents are of Native American ancestry.[21] Once encompassing more than 22,000 acres, Isle de Jean Charles is now only 320 acres.[22] More than 98 percent is underwater. It's our country's first community to receive federal funds as internally displaced climate refugees.

Another environmental injustice is something called climate gentrification, where poorer populations lose out to wealthier people as climate changes. Jesse M. Keenan, a researcher on urban development and climate adaptation at Harvard's Graduate School of Design, tracks three pathways here.[23] For one, as low-risk properties rise in value, less wealthy people are pushed out. For another, high-exposure properties become more and more expensive, pushing historically mixed-income areas like Miami Beach and the Hampton Roads to become exclusive. And finally, government investments in resilience boost property values, further displacing less wealthy.

Along the Atlantic Intracoastal Waterway 1,090 miles from Norfolk to Miami, sea-level rise problems magnify even more. Florida is a water state. With 1,350 miles of coastline, more than any other state besides Alaska, it's severely threatened by sea level rise as more than three-quarters of Florida's 22 million-plus people live along the coast. During the previous interglacial period 120,000 years ago, sea levels were twenty-five feet higher than today, with the Sunshine State a narrow stump of itself, stopping around St. Petersburg. When the last ice age peaked 18,000 to 20,000 years ago, lowering sea levels, the peninsula was almost twice today's width.

While all cities along Florida's coast will suffer in the coming decades, Miami is the most exposed city in the world in terms of property damage from climate change. Some $416 billion in assets are at risk, according to the Organization for Economic Co-operation and Development. And sprawling out further from the coast in South Florida, the odds do not improve much. Almost 60 percent of Miami-Dade County, which includes thirty-four municipalities, is less than six feet above sea level. Sea levels have risen nine inches there the past century. Optimists like the Southeast Florida Regional Climate Change

Compact, a partnership between Broward, Miami-Dade, Monroe, and Palm Beach counties,[24] expect another three to seven inches the next fifteen years—and nine inches to two feet over forty-five years.[25] Those are the optimists, mind you. We'll get to the pessimists in a minute.

Three threats from water exist in South Florida. Our attention in this chapter is on sea-level rise. As chapter 4 highlights, a second water-related threat is higher storm surges and more rainfall from higher intensity storms expected due to climate change. The third water threat in these parts is geological. South Florida sits atop a foundation of porous limestone up to one hundred feet thick. As opposed to granite or marble bedrock, which is essentially impervious to water, limestone is sediment that comprises millions of years of accumulated dead marine creatures like corals. These compressed ancient reefs, full of tiny holes, allow salty water to bubble up through the ground itself. "Our underlying geology is like Swiss cheese," notes Jayantha Obeysekera, a scientist with the South Florida Water Management District.[26] Dikes like those made famous in the Netherlands won't help South Florida because seawater here simply seeps in from underneath. "Being built on limestone, there's nothing you can do in the long run," agreed Jim Cason, senior advisor to the American Flood Coalition and former three-time mayor of Coral Gables. "Sea level rise is going to affect our aquifer underneath, and water will bubble up from below."[27]

It already is.

Water bubbles up through Miami storm drains as well as residential shower drains, sometimes mixed with the 22 million gallons of sewage residents create citywide each day.[28] There's a rather innocuous name for this, sunny day flooding. Tides are the culprit. As you know, these long-period waves slosh around the globe thanks to the gravitational pull of the moon and sun.

Twice a month, when Earth, moon, and sun orbits align during new and full moons, these tides are higher than normal. Earth scientists call these "spring tides," although it has nothing to do with the spring season. Twice a year, once in the spring and once in the fall (fall being the more extreme for us, when the Earth's northern hemisphere is closest to the sun), these spring tides become a perigean spring tide.[29] We call these king tides, the highest predicted high tides of the year.

With three feet of sea level rise, more than a third of southern Florida will disappear. At six feet, it's half. NOAA predicts up to 6.6 feet sea level rise by 2100. "Miami, as we know it today, is doomed," says Harold Wanless, a University of Miami professor of geological sciences. "It's not a question of if. It's a question of when."[30] The same can be said for iconic Miami Beach, once a bug-infested barrier island across Biscayne Bay. Early twentieth-century developers like Carl Fisher and John Collins transformed its mangroves and marshes into the international playground it is today. Miami Beach is also ground zero for sea level rise. A 2016 University of Miami study found that high-tide flooding in Miami Beach increased 400 percent since 2006.[31] This will only worsen as the U.S. Army Corps of Engineers, NASA, and the National Oceanic and Atmospheric Administration project between eight inches and six feet of sea-level rise by 2100.[32]

Wanless thinks these numbers are too low, suggesting that sea-level rise in Miami Beach and throughout South Florida will be much higher. Along with Philip Stoddard, then the mayor of South Miami, Wanless was named among Politico's top fifty "thinkers, doers and visionaries transforming American Politics in 2016."[33] He's also known around South Florida as Dr. Doom for his dire sea-level predictions. "If you look at seven feet by end of this century globally, you are looking at the end of South

Florida," warns Wanless. "There will be parts that are still sticking up, but infrastructure and connectivity will collapse. The risk from storm surge will become phenomenal."[34]

Miami Beach began spending $500 million for flood mitigation in February 2015, installing water pumps and raising city streets like those within the boutique neighborhood of Sunset Harbour two feet. But costs are now projected to top one billion and are forecast to provide benefits for only forty to fifty years. And those water pumps are greenhouse gas contributors themselves, meaning they might do more global harm than good in the long run.

Sea levels have changed throughout history, fluctuating more than five hundred feet. At the height of the last ice age, sea levels were three hundred feet lower than today. The previous peak, during our last warm period 125,000 years ago, when polar temperatures were only a few degrees higher than today, saw sea levels twenty to thirty feet higher.[35] But remember, the sea-level rise now unfolding will not be evenly distributed. Land is rising in Alaska due to glacial isostatic rebound. South Florida's higher projections are a product of lowered gravitational fields as glaciers melt and lose mass, as well as of the slowing Gulf Stream. "Those are not trivial," warns Wanless. "If there's a ten-foot rise in global sea level, those influences mean a fifteen to seventeen-foot rise here."[36]

With Florida's flat topography, that's going to eat up a lot of land. In fact, all along the East Coast, a gentle rise from the shoreline as you move inland translates into one to two hundred feet of land going underwater for every one-foot rise in sea level.[37] Wanless worries about what this means. He agrees with James Hansen that most models, which top out around six feet, fail to account for how rapidly our ice sheets will melt.[38] "The IPCC gave a really low projection because they're not incorporating

Greenland or Antarctica," says Wanless. "In 2008, we did not know that Antarctica had come alive. Now we know it's terribly alive, melting from below. And it's starting to get rainfall melting it from above."[39]

The day after our chat, Wanless published an article in *The Guardian* that spells out further details. "The US National Oceanic and Atmospheric Administration projected in 2017 that global mean sea level could rise five to 8.2 feet by 2100," Wanless wrote. "Four years later, it's clear that eight feet is in fact a moderate projection. And regional influences—subsidence, changing ocean currents, and redistribution of Earth's mass as ice melts—will cause some local sea level rise to be 20 to 70 percent higher than global. . . . Absent extensive and very expensive adaptation measures, it would put much of New York and Washington DC, Shanghai and Bangkok, Lagos, Alexandria, and countless other coastal cities underwater. It would submerge south Florida."[40]

Those living along the coast have a front-row seat to this continuing sea-level rise. But conscientious travelers also become better informed about such climate threats, from Hampton Roads in Virginia to the coastlines of Florida. That's the first step in addressing climate change: paying better attention to the threats it presents.

2

FLOODING IN THE FORECAST

... FLASH FLOOD WATCH IN EFFECT FROM 7 P.M. CDT THIS EVENING THROUGH THURSDAY AFTERNOON ...

The all-caps National Weather Service alert captures my attention. I'm in Texas Hill Country, what's known as Flash Flood Alley, fifteen miles north of the state capital of Austin, in a city called Round Rock. One of the fastest-growing municipalities in our nation, it's a community that has always worked around water. Just downhill from my motel is the town's namesake, a large, circular stone that nineteenth-century wagons once used to negotiate Brushy Creek. You might say deference to water is in Round Rock's DNA.

The next morning, speaking with city engineer and floodplain administrator Danny Halden, I learn that because of climate change, that's true now more than ever. "We're getting more rain than what previous data showed," asserts Halden. "FEMA five-hundred-year floodplain maps are really one-hundred-year floodplains if you throw in our new rainfall data."[1] With temperatures rising across this land and more intense rain events as a result, flood maps like those in Round Rock will require redrawing. Hundred-year floods could become annual events in some places like New England as coastal flooding from rising sea levels combines with stronger, more frequent storms.[2] Too much water isn't just a threat along our seashores, though. Flooding is becoming more and more problematic inland as well, from Buffalo and Pittsburgh to Louisville, Memphis, and Nashville.[3]

Overdevelopment is partly to blame, particularly if fast-growing locales like Round Rock fail to consider widening floodplains. But our changing climate is a root cause, enlarging floodplains as warmer temperatures absorb more moisture. It might not rain more often, but when it does, those rains are more and more intense. That's a problem. "We design our infrastructure and plan our society looking backwards, assuming that the past is a reliable predictor for the future," climate scientist Katharine Hayhoe explains. "And looking backwards does keep us safe, when climate is relatively stable, as it has been over much of the history of human civilization. When climate is changing, though, relying on the past to predict the future will give us the wrong answer . . . and a potentially dangerous one."[4]

What qualifies as dangerous might surprise you. As little as eighteen inches of water can lift a car off pavement and sweep it downstream. Eighty percent of flood deaths come from driving into such flooded roadways or people attempting to walk through this moving water, according to the National Weather Service. Beyond lives lost, floods caused $3.75 billion in property and crop damage across the United States in 2019.[5] California, South Dakota, Virginia, and West Virginia all experience more than their share of flooding losses. But the Lone Star State holds the most cities with homes susceptible to flooding, with Texas Hill Country serving as ground zero. "We have some of the highest rainfall intensities in North America," says Matt Hollon, division manager of planning for the Watershed Protection Department, City of Austin. "This is flash flood alley. We get the most rain in the shortest time."[6] "We did a climate resiliency study of city infrastructure three years ago, buckling roads, heat, fire, floods, anything we could think of," adds his colleague Tom Ennis, who leads Austin's Watershed Protection Department's sustainability efforts, including its Climate Protection Program. "In the end, eight out of the ten most vulnerable situations for the city of Austin were flooding. And that was all department's giving their two cents worth, not just our watershed department."[7]

Each spring, cool, dry air arrives here from the Rocky Mountains, where it meets warm moisture from the Gulf of Mexico. The resulting rain can fall so fast that the rocky, clay-rich Central Texas soils fail to absorb it. In the last decade alone, three hundred-year floods hit Central Texas. The Halloween Flood of 2013 unleashed nine to ten inches across Hays and Travis Counties. The Memorial Weekend Flood of 2015 dropped ten to thirteen inches in southern Blanco County, raising the Blanco River as astonishing thirty-six feet in only four hours.[8] And the Hill Country Flood of 2018 released eight to twelve inches of rain.

Even dry parts of the state have issues with flooding. The next day, four hours west in Abilene, I learn that firsthand. Abilene is considered the gateway to dry and flat West Texas. Yet at least 30 to 35 percent of the city lies within a floodplain. "A lot of our town was built in the middle of a floodplain, especially in the '50s and '60s when not a whole lot of attention was paid to that, at least out here," explains Abilene City Manager Robert Hanna.[9] A creek system spreads across Abilene like a hand with fingers. Storm waters normally drain quickly into the soil and creeks, but during heavy rainfall, as the ground becomes saturated, a capacity issue arises. That's the case now as I stroll around a neighborhood off Catclaw Creek, one of several watery fingers meandering through the city. The day before I arrived, a little over an inch and a half of rain fell. More than another half-inch dropped this morning.[10]

That's becoming more and more notable because, aside from a small section downtown, most of Abilene has no stormwater drains. "Streets serve a dual use here. We also use them as our drainage basin," explains Srinivas Valavala, stormwater services administrator for the city of Abilene. "That must stop at some point, though. With more intense rain, we'll have to accommodate changing climate patterns."[11] Rain the past two days may be indicative of that. Another downpour in this same neighborhood last month probably is, too. "There was just water everywhere," said one resident after water from Catclaw Creek flooded into her yard, stopping only inches from her front door. "It looked like I was in the middle of the lake. I thought my car was going to float away."[12]

People love living near water. Most of the world's great cities were founded along riverbanks or ocean shoreline. But be careful what you wish for. While living close to water has always carried flood risk, those risks increase as our hydrosphere changes. According to NOAA's National Centers for Environmental Information, from 1980 to mid-2021, 298 U.S. weather/climate disasters registered losses exceeding $1 billion each, including drought, flooding, freezes, severe storms, tropical cyclones, wildfires, and winter storms.[13] The cost of flood damage alone was approximately $17 billion annually between 2010 and 2018, according to Congressional testimony from Federal Emergency Management Agency representative Michael Grimm.[14]

Add to that costs that migrate from one city to another. "The number one state of climate migrants is going to be Florida. And one of largest cities that will be receiving those migrants will be the city of Austin," explains Ennis. "That will be taxing our social services. We have a hard enough time addressing social services with the population we currently have, almost doubling every ten years."[15] As Ennis notes, surviving a flood is only the first challenge victims face. The aftermath of floods inflicts more injury, disease, and death than the flood itself.[16] Floodwaters disable infrastructure and institutions, overrunning sanitation systems and contaminating water supply. Transportation and communication suffer. And loss of livelihood follows displacement. More than two-thirds of U.S. waterborne disease outbreaks come after unusually intense rainfall. A dramatic example was in Milwaukee in 1993, when 400,000 fell ill from a cryptosporidium outbreak immediately after a storm.

Major U.S. flooding costs have grown considerably the last couple decades. In 2019, the Missouri River flooded 61 million acres under a foot of water, enough to cover all New York, Connecticut, and New Jersey. Levees blew out or overtopped across three states, causing more than $3 billion in damages. An entire town was flooded. In 2011, record rainfall flooded the Mississippi River on a scale not seen since the Great Flood of 1927, forcing the Army Corps of Engineers to open spillways and blow a series of levees fifty-six miles long.[17] Thousands of square miles of agricultural and residential land flooded, with economic damages estimated at $2.8 billion. In 2010, the victim was the capital city of Tennessee, when the Cumberland River swamped Nashville in a thirty-six-hour deluge. That five-hundred to one-thousand-year flood set records in terms of water volume, property damage, and loss of life. Even the Grand Ole Opry House's main stage, home to America's longest-running continuous radio program, was submerged under four feet of water.

It started raining on Saturday morning, that first of May, intensifying throughout the day. Over the next day and a half,

Music City USA endured an unprecedented 13.57 inches of rainfall, more than doubling its previous record for a single rain event. "We didn't just barely edge out a record," Larry Vannozzi, meteorologist in charge of the National Weather Service's Nashville office. "We destroyed a record."[18] It should be noted that flooding is not new to Nashville. It's been a threat since settlers first arrived in 1779. A big reason for that is the Cumberland River, which winds for miles through metro Nashville along with several tributaries.[19] While the 2010 flood crested twelve feet above flood level at 51.86 feet, the highest since the Cumberland River dam system was built in late 1950s and early 1960s, it was not the highest ever recorded.[20] In any case, waters quickly overwhelmed much of the metro region. Eleven people perished within the Nashville area, and another fifteen died in other parts of Tennessee and Kentucky. Mayor Karl Dean declared a citywide state of emergency.

Rainy season in central Tennessee traditionally falls between December and April. I remember those wet winters and springs as a child, splashing around in the puddles they provided. But I never saw anything like this. It didn't help that almost three inches of rain fell the week prior to the storm, saturating soils. When the May 1 storm arrived, water had nowhere else to go. By the end of the weekend whole buildings, sheds, and barns, were racing down waterways like the Harpeth River, a Cumberland River tributary that flows through parts of Nashville and the southern suburb of Franklin. "There were lawn mowers, bicycles, tools, camping gear," said Michael Cain, director of watershed assessment and restoration at the nonprofit Harpeth River Watershed. "You name it, we found it!"[21]

Lieutenant Matt Pylkas, from the west precinct of the metro Nashville Police Department, received a Tennessee Homeland Security first responder award for his work along another part of the Harpeth River in West Nashville that weekend. Along with about 180 others, Pylkas was marooned on a temporary island in the middle of Interstate 40. Some were rescued by boats before sunset. Others ended up spending the night. It's an experience Pylkas says he will never forget. "Then night came, fog was rolling in," said Pylkas. "You could actually here animals drowning. . . . It was very quiet outside except for those distress calls of the cattle."[22] No part of the city was immune, from residential to industrial. Nearly 11,000 properties were damaged or destroyed in the flood. Some 10,000 people were displaced from their homes. The Nashville Chamber of Commerce reported 2,773 businesses impacted. Private property damage was estimated at $2 billion, with another $120 million in public infrastructure damage in Nashville alone. About 450 roads and bridges needed repair. Perhaps the most indelible image was a portable classroom floating down Interstate 24.[23]

One year after the flood, *The Tennessean*, citing the Nashville Area Chamber of Commerce, reported that 300 to 400 businesses remained closed and 1,528 jobs were "very unlikely" to return."[24] "The magnitude of it, the economic impact of that flood on this area was equal to a year's growth GDP," said Ralph Schulz, CEO of the Nashville Area Chamber of Commerce. "We just extended our recession another year. The water subsided over two or three days, but the damage was left behind."[25]

Climate change was not alone in complicating the 2010 Nashville flood. The U.S. Army Corps of Engineers and National Weather Service took heat in congressional hearings in the months that followed, the NWS for not communicating clearly enough the magnitude of the rain event and the Corps for releasing six inches of water from the Cordell Hull and Old Hickory dams in preparation for the storm. But Corps officials maintained that Old Hickory Dam may have overflown and the

Cumberland River would have crested four feet higher had the Corps not released that water. They add that the Old Hickory dam came within seven inches of reaching its maximum elevation and threatening to flood Corps facilities.[26]

That raises another reasonable question: Why were dams kept so high to begin with, particularly during the rainy season? The answer demonstrates why we must better understand how our daily consumption impacts local surroundings. Dam levels need to stay high, even though that lessens their storm capacity—if they are to provide sufficient electrical power.[27] You could argue, then, that flooding events like Nashville's were driven not only by fossil fuels but also by demand for hydropower. Becoming more aware of these relationships, seeing these local impacts yourself in your own communities, also hammers home the message that poor development decisions have consequences. "Old shopping malls and parking lots built in flood-prone areas can cause an overall rise in floodwaters that can damage other property owners who might not otherwise get flooded," notes Nashville Metro Councilman Jason Hilleman. "Often, it's the people near these over-paved developments who did nothing wrong and had no reason to think they were at risk who pay the price when the rain falls."[28]

Antiquated infrastructure also magnifies risks. In older cities like New York, for example, aging sewer systems connect with wastewater systems. During normal weather, all this water is treated effectively at wastewater plants before being discharged into bays and rivers. But during severe rainstorms, sewage treatment plants are overwhelmed, with untreated sewage released directly into waterways.

As we will discuss in part II of this book, sprawl is still another human factor. According to the *Water Resources Bulletin*, when 25 percent of natural terrain is covered in asphalt, concrete, and rooftops, a once-in-a-hundred-year flood can occur every five years. Texas Hill Country bears that out. Austin area communities like Round Rock are growing at incredible rates. Round Rock alone has grown from 34,000 in 1990 to more than 120,000 in 2020. Its landscape and quality of life make this an attractive place to live. The area prides itself as pro-business, with low taxes and low utility rates. But officials must be careful not to become a victim of their successes. Round Rock has the largest floodplain population increase of any city in our country between 2010 and 2016.[29] What does this mean when the next flood inevitably comes?

National initiatives like the National Flood Insurance Program (NFIP) further complicate matters, encouraging construction in areas where it should not occur. Created by Congress in 1968, the NFIP replaced underwriting by private insurance companies refusing to write flood policies they deemed too risky. It has lost $50 billion to date and is still $20.5 billion in debt, despite repeated taxpayer-funded bailouts.[30] When do we say enough? When do we recognize our hydrological cycle is changing, that a warmer atmosphere increases the intensity of precipitation events and the risk of flooding?[31]

Climate change dictates more forward-looking mapping and assessment of risks. Creating a more permeable cityscape will also help us adapt, a subject to which we return later in this book.[32] Even so, "We just can't engineer ourselves out of this problem," as Bill Hunt, a civil engineer at North Carolina State University, asserts.[33] We need to rethink where we build—as well as where we have already built. And we need to confront the root causes of more frequent and intense flooding, climate change.

3

DROUGHT AND WILDFIRE

It's 113 degrees as I pull into Paradise. The temperature more than doubled during my four-hour drive from San Francisco to this small northern Californian town. But triple-digit heat is not my only concern. It's also extremely dry, with only 9 percent humidity. Together, this heat and dryness are a worrisome welcoming mat for western wildfires. I see smoke from one burning now, about a hundred miles north. Smoke is a good indicator of how dangerous a wildfire is. Dark smoke comes from heavy fuels burning and spells trouble, while lighter-colored smoke is less concerning as its fire feeds off grasses and smaller vegetation. This one, reassuringly, is the latter, which calms the nerves of those familiar with Paradise.

The town sits in the Sierra Nevada foothills, on a wide ridge between deep canyons formed by Butte Creek on its west and the west branch of Feather River on its east. Chico, a college town of about 100,000, is twelve miles to the west. Paradise's origins trace to California's gold rush. One claim to fame is the 1859 discovery of a fifty-four-pound gold nugget, the largest ever found in North America. Another is the most lethal and costly wildfire in Californian history. Sparked by nearly hundred-year-old Pacific Gas & Electric (PG&E) equipment and strong winds that dropped an electrified line into dry vegetation, the November 2018 Camp Fire killed eighty-five and turned this community of 26,543 into a burned shadow of its former self.

Perhaps six thousand people now live in Paradise, although that generous figure likely includes temporary workers helping rebuild. My last four days of interviews have been eye-opening, a testament to the resiliency of those that returned to this ridge—but also a powerful reminder that our climate is changing, making wildfires an ever-increasing threat. As I prepare for the monthly town council meeting, I receive a text from former council member Melissa Schuster: "So, this is happening . . ." There's another fire, one eerily close to where the monster Camp Fire first ignited. This Dixie Fire looks not to threaten Paradise, though. Wind, a critical third element to fire, is in the town's favor this time, and over the next twenty-four hours it pushes flames northeast toward less populated areas.[1] Yet, in the weeks to come, this Dixie Fire will become the state's second largest wildfire ever, burning nearly a million acres.

This land is a beautiful part of our country. People love living here. But climate change is making that more and more dangerous. Hotter temperatures are soaking up more moisture and prolonging California's current drought, which increases chances for wildfire. According to the U.S. Forest Service, fire season is seventy-eight days longer than fifty years ago.[2]

Large wildfires are named after the town, region, or watershed where they ignite. The Camp Fire started along Camp Creek Road, a narrow stretch of dirt 7.5 miles east of Paradise in the unincorporated community of Pulga. Initially, it was only a small canyon brush fire. But Camp Creek Road is hard to reach. And strong winds complicated matters quickly. Gusts over 50 mph prevented flying in to fight the fire. Embers began blowing everywhere. At peak spread, the fire advanced an astounding eighty football fields a minute.[3] Paradise, along with surrounding communities totaling over forty thousand inhabitants, had little chance.

"It looked like Mordor, a black inky streak in the sky," said David Leon Zink, board chair of the nonprofit Regenerating Paradise.[4] That foreboding dark smoke sign soon worsened. Day turned into night. "My first thought was, oh, thank goodness it's raining," said Schuster. "But it wasn't raining. It was ash. And then there was the darkness. You couldn't see four feet from me to you. I'd never experienced anything like that."[5] California's deadliest, most destructive fire on record reduced human remains to mere bone fragments. It took weeks, using DNA analysis, to identify many of those who perished. The fire kept burning for seventeen days, until winter rains finally came. But it took only four hours to sweep through Paradise, destroying 18,804 structures. All total, the Camp Fire claimed 153,336 acres, an area larger than the city of Chicago. "It looked like a war zone, like a moonscape," described Bill Hartley, a Paradise resident since 1982.[6]

Two previous fires hit parts of Paradise a decade earlier, providing some warning. "God was ringing the doorbell," noted longtime native and Golden Nugget Museum Board President Don Criswell. "We just weren't answering the door."[7] The Humboldt Fire burned 22,800 acres between Chico and Paradise in June 2008, forcing the southwestern part of town to evacuate. Just weeks later the northeastern side of Paradise was threatened and thousands there evacuated, although, thankfully, that fire failed to cross the river in the Feather River canyon.

The Camp Fire was different. "What made it different was how it spread, how it spotted," explains Joe Tapia, Cal Fire Battalion Chief for Butte County, as his radio crackles in the background. "There wasn't a big flaming front. It was everywhere—and kept leaping ahead."[8] Chief Tapia has seen his share of fires, starting as a seasonal firefighter back in 1989. "The fire was two miles wide when it hit town," he elaborates. "But the wind was blowing really hard, so there were spot fires everywhere. Most of fire spread was by ember. Sometimes just a blizzard of embers. Instead of snowflakes, it was burning embers. And houses were the fuel."[9] Paradise's disaster recovery director, Collette Curtis, agrees. "The Camp Fire didn't behave in the way fires have in the past," Curtis says from her city hall office. "Normally there's a fire front, and you create containment around it. But this one was spotting miles ahead. That's how Paradise caught fire so quickly. That type of fire behavior is something we hadn't seen before."[10] Paradise residents had little warning as a result. The typical day or two notice shrank to mere minutes.

In our changing climate, with enhanced risks like these, it's easy to forget that traditional wildfires are both natural and necessary, bringing multiple benefits to ecosystems. Some tree species disperse seeds only during fire. Lodgepole pines, for example, have heavy resin seals on their pinecone scales that only break apart to release their seeds at high temperatures. Wildfires also clear out disease and insects amidst fallen timber while returning essential nutrients to the soil. That promotes grass growth for elk and bison as well as ground-foraging birds and small mammals who feed on the seeds of those grasses. Those birds and small mammals, in turn, feed hawks, owls, and coyotes. Native Americans understood these benefits and used them to their advantage, periodically starting fires themselves to attract game and promote berry growth.

But for over a century, we chose a very different path. The U.S. Forest Service, established in 1905, initially focused on logging and sheep grazing.[11] Fire suppression became a focal point after the Great Fire of 1910 burned an area the size of Connecticut, three million acres in in Washington, Idaho, and Montana. The Forest Service then shifted to a policy of containment and extinguishing, one that ramped up considerably in the 1940s as generations of Americans were raised on Smokey the Bear's memorable warning that, "Remember . . . Only YOU can Prevent Forest Fires."

Alas, instead of putting out fires, we've merely been putting them off, as author Philip Connors points out, drawing on eight seasons as a fire lookout.[12] A "managed wilderness" imposes artificial distinctions between human and nonhuman. This attempt to control our forests disrupts ecological interdependence. Over time, fire suppression leads to an increase in flammable undergrowth, paradoxically providing the ideal staging ground for larger, less controllable fires. Forests will burn. The only question is how and when. That said, wildfires are not the same everywhere. Western forests, with their spruce, pine, and larch trees burn more easily than deciduous trees such as oaks mixed with pine and maples in southeastern forests like the Great Smoky Mountains. It's also wetter in the eastern part of our country, so dead trees rot away more quickly than those in the drier West.

While wildfire is an age-old threat out west, from California to the Rockies to Alaska, now those threats come sooner and last longer. Even Alaska's peatlands burn. Peatlands are partly decomposed organic matter covered in sphagnum mosses. Historically, this waterlogged moss resists fire, but longer hot spells are drying it out, making it fire prone. Once they're lighted, peat fires can smolder for months. With only 733,000 people spread over a large territory, though, Alaska does not always capture our attention in terms of fire risk. California, with 39 million people in less than half the space, does. Not coincidentally, the

Golden State is also our country's richest state as well as our largest agricultural producer.[13] Much of the water supplying this agriculture comes from the nearby Sierra Nevada mountains. Indeed, as much as 65 percent of the state's total water supply comes from snowpack.[14] But climate change is shutting down this water storage service and further drying out this landscape.

Over the last five decades, our western wildfire season lengthened two and half months, regularly extending well into November. The science here is relatively simple. Just as we saw in the previous chapter, warmer air holds more water. It evaporates more moisture from the land, too, drying it out. But it's not just higher temperatures soaking up more moisture that feeds these more frequent and more intense fires. Additional plant growth during the rainy season dries out earlier now, providing added fuel when fires do alight. "That type of fire activity really wasn't seen here until much later in the summer," explains Curtis. "This year the conditions all around us are so dry and so hot it's more like August in June."[15]

Climate change and ill-conceived fire suppression during the twentieth century are not the only human factors at work. Overdevelopment, more specifically building on the forest edge, even within it, also increases the risks of wildfire. Known as the wildland-urban interface (WUI), this describes development within or at the border of wildland vegetation. Some 38.5 percent of American homes nationwide fall within WUI. In nineteen states, it's over half. And more people are moving into these fire prone areas every year.[16] "This ridge was ready to burn and had been for decades," longtime Paradise resident Don Criswell asserts. "Don't get me wrong. I'm no fan of PG&E, but it could've just as easily been a lightning strike that caused the same result."[17] Criswell moved back to Paradise in 1998 and speaks fondly of where he grew up. The next day he invites me out to his hundred-acre ranch, surrounded by another thousand plus acres of Bureau of Land Management territory. "Grazing is probably all this land ever should have been used for," he admits. "I'm one of those guys that moved into the wildland-urban interface. We've paid the price for that twice now."[18] But Criswell and his wife Debbie were also forward-thinking. The couple weathered the Camp Fire, fighting seventeen hours straight to save their home and barn. That was possible only because they created a perimeter on their property, a blank swath of land that limited the fire's ability to spread.

People technically start 84 percent of U.S. wildfires, accounting for nearly half of what burns.[19] Culprits run the gambit, from cigarettes tossed out of windows, improperly attended campfires, and brush and debris burns gone astray to fireworks, loose sparks from weed eaters and other machinery, downed power transmission lines, and arson. Even a gender-reveal party, with its smoke-generating pyrotechnic device, set off a 2020 wildfire that burned thousands of acres east of Los Angeles.[20] And climate change remains an underlying factor in many of these cases, even when not supplying the initial spark. Of course, with warmer temperatures bringing more lightning, climate change will undoubtedly assume a greater role in igniting wildfires, too. Approximately 44 percent of western wildfires between 1992 and 2015 stem from lightning, according to the U.S. Forest Service's wildfire database, and those fires were responsible for 71 percent of areas burned.[21] As this land becomes drier, and with researchers predicting a 12 percent uptick in the number of strikes for every 1.8 degrees Fahrenheit rise in temperature, those numbers will inevitably rise. Wind, also shaped by climate change, is another constant worry.[22] Wind feeds additional oxygen to a fire, increasing its intensity. And strong winds can send embers miles ahead of the main burn. In the case of Paradise, wind "spotted" fires as much as five miles ahead.

Mike Flannigan and Charlie Van Wagner warned of a wildfire connection to climate change three decades ago, suggesting a 46 percent increase in fire severity if temperatures were to rise as expected.[23] More recently, John Abatzoglou and Park Williams have found that wildfires out West burned almost double the area they would have if not for climate change.[24] Climate change is making it hotter, drier, and windier. Heat dome effects like this week, with prolonged triple digit temperatures, are a case in point. So is this continuing drought. While annual precipitation across most the northern and eastern United States has increased since the beginning of the last century, it decreased across much of the southern and western United States, according to our last National Climate Assessment.[25] Sometimes it's eight months between rains. Drought has always been a California climate feature, but climate change now brings them more frequently and more severely. Dry extremes previously seen on the order of every 150 years will likely occur every four, as often as the Summer Olympics. The simple fact is wildfires are more frequent today because of climate change. UC Berkley climate scientist Patrick Gonzalez estimates, since 1984, climate change caused twice as many wildfires as would have normally been expected.

Wildfires are also bigger. Size depends on five key variables. One is how old and expansive a forest is. Another is what natural breaks and barriers exist. The last three are the trifecta of climate change: how hot it is, how dry it is, and how fast the wind is blowing. The biggest fires are called megafires, loosely defined as burns of 100,000 acres or more. They can entail heat so hot it sterilizes soil, so vegetation struggles to grow back. Fewer than

3 percent of the fires in North America are megafires, but they now account for over 90 percent of areas burned.[26]

Climate change is also creating some troubling feedback loops. U.S. forests from 1990 to 2015 sequestered approximately 11 percent of our nation's carbon dioxide emissions.[27] But when trees die, they release their stored carbon back into the atmosphere. A single California wildfire can cancel out a year's worth of emissions reductions from this state's progressive environmental policies.[28] Globally, deforestation causes 12 percent of greenhouse gas emissions, with forest fires responsible for 25 percent of that amount. The loss of trees, of course, also means less carbon is stored, which translates into more warming and more drought. That, in turn, kills still more trees and feeds future fires. The increasing frequency of fire also creates a feedback loop. Burn-scarred landscapes are more susceptible to invasive plants, which tend to be more flammable and ignite more often. The El Niño Southern Oscillation (ENSO) creates a similar dynamic. The Pacific Northwest and California see increased winter precipitation during this cycle, every seven to ten years. That translates into a less risky fire season during El Niño years. But it also encourages greater vegetation growth, and in the warmer, dryer La Niña years that follow, these plants typically dry out and fuel even bigger fires.

Still another feedback loop revolves around an earlier spring that reduces snowpack. Less snow combines with earlier snowmelt to provide less water to rivers later in the year when it's hottest. That means dams on those rivers produce less clean hydroelectric power, which often translates into more coal produced power and more greenhouse gas emissions to push temperatures even higher. On top of that, less snowmelt and runoff mean farmers must install bigger electric pumps to bring water up to the surface for irrigation. That means greater electricity demand, which, if the power is supplied by fossil fuels, adds still another feedback loop to climate change.

Finally, the mountain pine beetle is another potential feedback loop, killing trees that can fuel still more fires. A tiny, tree-boring insect the size of a grain of rice, it is one of several insects damaging North America's coniferous forests like lodgepole pine and jack pine. Mountain pine beetles live one year, laying their eggs, up to seventy-five per pair, inside trees during the summer months. Historically, most of these larvae die off during winter, with only two percent surviving. But that is changing as winters become warmer and shorter. Without extended frigid weather, more and more beetle larvae survive. Exactly how cold and how long that cold must remain to retain ecological balance is not fully understood because beetle larvae employ alcohols that protect them from freezing. These antifreeze levels are lower in early fall and late spring, making them more susceptible to cold temperatures then, but studies show that temperatures from –13° F to –31° F in midwinter can cause mortality, too, according to the U.S. Forest Service's Rocky Mountain Research Station.[29]

Costs from drought and wildfire are many—and high. A major wildfire season like 2015 can release ten years' worth of stored carbon in a single year. On top of that, as noted earlier, we lose a valuable carbon sink when those trees are lost. Wildfires are not cheap to fight, either. The U.S. Forest Service devoted more than half its appropriated budget there in 2015. That's up from 16 percent in 1995, meaning significantly less money available for forest management and maintaining recreation. And then there are the impacts after the burn. Heat-damaged plastics contaminate water infrastructure, leaching dangerous chemicals like benzene. The EPA classifies water with benzene levels above five hundred parts per billion as hazardous. Samples in Paradise registered two thousand parts per billion.[30]

It's not just the local environment that suffers. In the right conditions, smoke and ash can travel hundreds, even thousands of miles. Fires in the Northwest Territories of Canada in 2014 affected air quality in Chicago and New York. In 2013, ash from a Canadian wildfire darkened portions of the Greenland ice cap.[31] That further increases the rate of climate change because, instead of being reflective, previously white glaciers now covered with darker ash absorb more sunlight and melt the ice even quicker.

There are also agricultural impacts to consider. A severe Midwestern and Western drought in 2012 raised food prices 6 percent globally, with corn alone rising 23 percent.[32] States like Oklahoma and Texas have much to lose, including nearly 25 percent of their cattle during a record 2011 drought. But California, with over 76,000 farms and ranches, holds the biggest farm economy of all our states. It is our leading producer of vegetables, fruits, nuts, and dairy products, supplying over a third of the country's vegetables and two-thirds of its fruits and nuts. With changes in temperature and moisture thanks to climate change, growing seasons here will be markedly different. Warmer temperatures, with more time between frosts, can translate into a longer growing season, if those increases are slight. But lack of moisture, and increasingly hot temperatures, more than cancel those benefits out. Researchers at the University of California predict rising temperatures will reduce avocado yields 40 percent by 2050, with almond, table grape, orange, and walnut production dropping 20 percent.[33] Our changing climate may make 54 to 77 percent of California's famed Central Valley unsuitable for apricot, kiwifruit, peach, nectarine, plum, and walnut farming by the end of this century, namely because fruit and nut trees have "chilling requirements" where they need a certain number of cold hours for their buds to go dormant and reset for spring production.[34]

Three years after the Camp Fire, Paradise is still rebuilding. It will be for years. But within this tragedy new opportunities arise. "The gift that the Camp Fire gave us is that it helped us to see what isn't working. People became permeable to new ideas," says Wolfy Rougle, forest health watershed coordinator with the Butte County Resource Conservation District.[35] More specifically, the Camp Fire burn scar offers insight into how forests might better withstand climate change. Forests better resist fire when they have fewer trees. It also helps if those trees are larger and more widely spaced. Of course, that only goes so far if Paradise doesn't make additional changes. Four evacuation routes were not enough to empty the entire town. With a road network rooted in former gold mine trails and apple orchard paths to maximize buildable space, Paradise features far too many dead-end streets and minimal connector options.[36] A road diet, narrowing a stretch of Skyway Road in the heart of downtown to slow speeding traffic four years before the fire, also received plenty of criticism.

But the root problem, along with climate change, is overdevelopment. Too many people were on the ridge that fateful morning. And Paradise was set within a forest, lacking something called defensible space. California's Public Resources Code Section 4291 mandates a hundred-foot perimeter of "defensible space" from flammable materials. That requirement, though, ends at the property line, and Paradise homes were often built beside one another within that perimeter. Some are rebuilding now within that same flawed footprint. Some already have, even installing wooden fences that previously acted as wicks linking one burning house to another. Paradise did propose some twenty changes to its building code, with nine adopted just over half a year after the fire. Still, more than half of the recommendations were not passed—and more than half of those that did were modified after public objections. One ordinance that no combustible material be within five feet of a home was adapted to allow certain plants. Another ordinance prohibiting gutters unless over entranceways or preventing erosion was changed to allow noncombustible gutters anywhere, ignoring that all gutters collect highly combustible pine needles.

Those critiques aside, officials like Paradise Recreation and Parks District Manager Dan Efseaff understand the stakes. Teaming up with the Nature Conservancy, Efseaff hopes to construct a defensible buffer zone around the town, one that doubles as a recreation area. This green firebreak will allow firefighters to carry out controlled burns, decreasing the chances large wildfires break out.[37] "That gives us a chance to maintain it, to actually do something before the fire rather than hoping for the perfect set of conditions to combat a fire after it's started," explains Efseaff.[38] Local actions like this matter, especially with the changing nature of wildfires owing to climate change. Californians understand this better than many Americans, in large part because they see it play out in their own backyards annually.[39] But the rest of us can also learn from the Golden State and other locales that face drought and wildfire. This is not an argument for disaster tourism. Neither is it one that ignores greenhouse gas emissions come with all travel. That said, how we travel matters. And what we do after those travels does, too.

4

MORE EXTREME WEATHER

An alarm blasts. Bad storms are brewing as ever-darkening skies reinforce my melancholy mood. I dropped my daughter off for her first year of college this weekend and am slowly absorbing what that really means. Many parents know this day will come, but, in contrast to my thirteen-hour return drive home, her childhood seems to have passed much too quickly. Sheets of rain distract me for now, somehow becoming ever stronger each moment. The National Weather Service announces a tornado warning for the next half-hour, and I ponder my limited options along this isolated stretch of interstate.

The next day I learn more details about this close call. An EF2 tornado, with a funnel cloud five hundred yards wide, touched down roughly five miles from my location, damaging two dozen structures in Deland, Florida.[1] The power of such a storm stirs up a mixture of emotions from awe to fear to despair. Now imagine a tornado six times wider at nearly two miles across and several orders of magnitude stronger—with winds an astounding 205 mph. On top of all that, instead of swooping quickly through, that tornado sits upon an entire town for eight and a half minutes. Greensburg, Kansas, population 1,500, faced such a storm one Friday evening in early May 2007. Ninety-five percent of it was destroyed.

The first EF5 on record, with a defined eye like a hurricane, that tornado obliterated 961 homes, extensively damaging another 216 structures while 307 more sustained minor damage.[2] The courthouse, at least its basement, was the only public building left standing. Yet, amazingly, only eleven people died. "The sirens started shortly after 9:15 that night, so people heeded that warning," explains five-time former Greensburg mayor Box Dixson. "People hadn't gone to bed yet. If they had, it could have been well over one hundred. The National Weather Service probably saved at least that many lives."[3]

Indeed, other tornadoes over the last decade or so have been much more deadly. Sixty-four died when an EF4 hit Tuscaloosa, Alabama, in April 2011, while 162 perished from an EF5 striking Joplin, Missouri, less than a month later that May. And almost a century ago, the infamous "Tri-State Tornado," covering more than three hundred miles in Missouri, Illinois, and Indiana, still holds the record as our nation's deadliest tornado, killing 695 and injuring 2,027 in March 1925. Greensburg's tragedy, then, could have been much worse, particularly since its storm holds the unwanted distinction of being the first EF5 tornado after the Enhanced Fujita scale was adopted only three months earlier.[4]

Established in 1886, the town lies about 110 miles west of Wichita astride U.S. Highway 54, approximating the route of nineteenth-century stagecoach driver and town namesake Donald R. "Cannonball" Green. It's also in the middle of what has been historically known as Tornado Alley, a broad north–south stretch of land cutting across South Dakota, Nebraska, Kansas, Oklahoma, and Texas. Alas, nowadays the term *alley* fails to accurately describe tornado territory. *Blob* is more accurate.

Meteorologists increasingly find twisters just as common, and often more deadly, in the South and Southeast from Mississippi and Alabama to Georgia and Tennessee. "The term 'alley' is restrictive, suggests something that is spatially long and narrow," P. Grady Dixon, a physical geographer at Fort Hays State University in Kansas, told a reporter. "We have a tornado region that's essentially the eastern 40 percent of the continental U.S."[5]

Here on the plains that threat begins when the wet season and warming air coincide during late spring, with my mid-April visit now through June serving as peak season.[6] And that combination of wet and warm, along with wind, barometric pressure dropping, and hail, tore this small western Kansas prairie town apart. "Bad weather was brewing all afternoon," recounts Greensburg School Superintendent Staci Derstein, then the elementary school principal. "We cut short a high school golf tournament with my son about twenty-five miles south to return home. Along the way, all these stormchasers were passing us. When we got home, the Wichita weatherman was on TV shaking, his voice quivering as he talked about the storm, calling it a rain-wrapped tornado."[7] That made an impression on Derstein, convincing her that this storm would not be one to take lightly. Later that evening, so did massive hail, as big as apples and softballs, as she remembers. "And then there was this weird pressure drop and a whoosh," recalls Greensburg resident Shawn Cannon. "It sounded like a wrecking ball pounding away on our home."[8]

That's a common memory among those in Greensburg that evening, or anyone who lives through any tornado, for that matter: the deafening roar of wind and debris, comparable to a freight train bearing down on you and all your worldly possessions. But former Mayor Bob Dixson remembers another eerie sound as he reminisces about his historic 1912 Victorian home. "If you've ever tried to pull a nail out of an old yellow pine with a crowbar, it just creaks and groans," says Dixson. "Multiply that by ten thousand nails. That's what I remember hearing."[9]

Then, immediately after a tornado, there is a heavy, dark silence. All power is lost, so the only light comes from lightning flickering in the distance. The air is full of grimy smoke you can both smell and see. And debris is littered everywhere, in pieces both large and small. "The second half of the tornado we had water coming through our floorboard to our basement shelter, so I knew it wasn't good upstairs," recalls Derstein. "The east wall of our house had ripped off, but other walls were oddly still there with furniture sitting in place. It looked like a doll house. Some rooms still had stuff hanging on walls, others nothing left."[10]

Nothing left is a frequent refrain. Current Greensburg Mayor Matt Christenson was a graduate student in software engineering at Kansas State when the tornado hit. He returned to help his parents rebuild as soon as the National Guard let folks back into town three days later. "I didn't recognize a thing," he says. "All the landmarks and buildings, trees I'd seen all my life, they weren't there anymore. I didn't know where I was."[11]

All this is hard to imagine for an outsider like me as I stroll through the wide, wind-swept streets of Greensburg almost fourteen years later, taking stock of its recovery efforts. Vacant, corner lots with exposed home foundations are a vivid reminder. Cracked, grass-covered sidewalks leading past these neighborhood gaps are another. And then there are the trees. Hundreds were lost in the storm, so the lack of older, established species throughout town is striking. But perhaps even more conspicuous are the trees that survived, scraggly and stripped of their former majestic canopy. They serve as a continual, almost ghostly reminder. Again, other cities have experienced more loss in terms of total numbers. Measuring tree canopy, for example, three hurricanes swept through my hometown of Winter Park,

Florida, in fall 2004,[12] destroying at least eight thousand trees, including almost a third of city-owned rights of way and park trees.[13] But Greensburg lost a much higher percentage, with those that survived only a shadow of their former selves.

Many residents vowed to overcome these obstacles, returning to rebuild in the weeks and months that followed. But many also moved on, choosing to start over somewhere else. Greensburg, like innumerable small towns across our nation, struggled well before this storm, with high school graduates serving as the town's largest export. Some feared this EF5 tornado would sound a death knell. And Greensburg, indeed, struggled, with the 2020 Census listing a population of only 772, almost half its pre-storm number of 1,500.

Natural tragedies worldwide can force such migrations. Some twenty million people become homeless each year across the globe due to floods, storms, and other disasters. That's roughly three times as many as those displaced by conflict or violence.[14] Even wealthy countries like ours are not immune. And that was true far before climate change became a prominent factor. Our nation's worst natural disaster in terms of fatalities, for example, fell over a century ago in September 1900. More than eight thousand perished in Galveston, Texas, then, according to the National Hurricane Center. Still, as illustrated with the Camp Fire in Paradise, California, in the previous chapter, extreme weather is more and more likely as our climate becomes warmer. No single storm, whether hurricane or tornado, can be attributed to climate change. But warmer air holds more water, leading to moister and less stable atmospheric conditions that create the summer thunderheads conducive to tornadoes. This is called convective available potential energy, or CAPE for short.

Climate scientists are not certain this means more tornadoes. Twisting winds also factor into tornado formation, and the wind shear that creates this condition, when wind speeds shift or change direction or height, is driven by differences in temperature. Some climate scientists suggest CAPE will supersede changes in wind shear, causing an increase in tornado intensity. Others contend, even as CAPE increases, that less wind shear will translate into little, if any, change in total tornado activity.[15] But the common concern is storms of all stripes could be stronger, if not more frequent. Marshall Shephard, director of Atmospheric Sciences at the University of Georgia, draws perhaps the best analogy I've heard. "You can grow grass almost anywhere," Shephard explained one October morning to a group of climate activists at the University of Central Florida in Orlando. "But when you put fertilizer on it, that grass really blossoms. In short, climate change is fertilizing more extreme weather."[16]

Again, to be clear, that doesn't necessarily mean tornadoes or hurricanes will form more often. But when these storms do develop, they will wield more wallop. And across the country, from intense rain to acute drought, climate change clearly brings more extreme weather, even without considering tornadoes and hurricanes. According to the last National Climate Assessment, based on reporting by scientists from thirteen federal agencies every four years, each corner of our country faces this fact. Average annual rainfall increased about 4 percent since the beginning of the twentieth century. But this is an average unevenly distributed across our land. The southern plains and Midwest are getting wetter while parts of our West, Southwest, and Southeast dry up. Our weather, as D. Hunter Lovins, cofounder of the Rocky Mountain Institute, suggested over a decade ago, is becoming weirder.[17]

Moving from the Midwest to the coast, particularly the Gulf Coast, tropical storms are another case in point. True to their name, tropical storms originate out of the warm, moist tropics and require waters of at least 80 degrees Fahrenheit to develop. Seven global hot spots exist: three around the Indian Ocean, two in the North Pacific, one in the southwestern Pacific, and one in the northern Atlantic.[18] Tropical storms that reach 74 mph graduate to hurricane status, forming their famous central eye around 80 mph, and, at times, travel thousands of square miles.[19] Strength is measured by low pressure at that eye using the Saffir-Simpson scale created in 1971 by Miami engineer Herbert Saffir and meteorologist Robert Simpson, who led the U.S. National Hurricane Center (NHC). Like tornadoes, this ranking uses a scale of one to five, with the highest Category 5 storms sustaining "catastrophic" winds of at least 157 mph (250 km/h). This means, in the phrasing of the National Hurricane Center, that "a high percentage of framed homes will be destroyed, with total roof failure and wall collapse. Fallen trees and power poles will isolate residential areas. Power outages will last for weeks to possibly months. Most of the area will be uninhabitable for weeks or months."[20]

But wind is not the only factor. Barometric pressure and storm surge are also critical. Flooding from the storm surge with the 1900 Galveston Hurricane is a case in point. Storm size and how fast a storm moves matter as well. Hurricane Dorian stalled over the Bahamas in September 2019, for example, unleashing its wrath of wind and storm surge for nearly a day and a half. The topography of both the ocean bottom and the shoreline where a storm makes landfall also affects how destructive hurricanes become. That was certainly the case in August 2005, with Hurricane Katrina hitting low-lying New Orleans even as its wind speeds dropped shortly before landfall.

There have been thirty-five documented hurricanes in the North Atlantic that reached Category 5 status since 1924. Of those, four have hit the United States at full Cat 5 strength, most recently Hurricane Michael in 2018. Michael was a tropical storm only a few days prior to its landfall but rapidly gained in strength to sustained winds of 160 mph and brought catastrophic storm surge to the Florida Panhandle along with extensive wind damage considerably inland, from the panhandle into southwest Georgia.[21] Kerry Emmanuel, a theoretical meteorologist at MIT, predicted intensifying hurricanes like this due to climate change back in the mid-1980s. Decades later his research demonstrates their increasing destructiveness.[22]

To be fair, other factors make destruction from extreme weather worse, too. In Louisiana, the loss of wetlands means less protection from storm surge. Development is a primary driver here, including, ironically, canals built by the oil and gas industry so that they could more easily transport drilling rigs and equipment needed to install pipelines along the coast to distribute those greenhouse gas–emitting fuels. In Mississippi, the legalization of the gaming industry in the 1990s spurred additional development and significant population migration to the coast. More people moved into harm's way. And in Florida, as Pulitzer Prize–winning author Jack Davis writes, an insatiable desire for beachfront property drove a century-long assault on mangroves, destroying not only the mangrove trees themselves but also vast swaths of "vegetation and the peaty soil around their tangled feet, [ignoring the fact that] mangrove forests absorb more carbon dioxide . . . than any tropical forest."[23]

Such development practices and their wanton disregard for our environment cannot explain other troubling statistics, though. In the fourteen years between 1975 and 1989, 171 storms were classified as Category 4 or 5. In the same timespan from 1990 to 2004, that number rose to 269. In the North Atlantic alone, the number of Category 5 hurricanes tripled.[24] None can be blamed definitively on climate change, but a warmer planet makes such extreme weather more likely. Take the 2020 hurricane season as an example. It spawned a record number of named storms in the Atlantic Ocean, moving through the

Greek alphabet for the first time since 2005 as thirteen of the thirty storms became hurricanes—six of those major ones.[25] Although we focus our attention in the states on the Atlantic, the trend is even worse elsewhere. The largest increase in absolute number as well as proportion of hurricanes reaching Categories 4 and 5 is in the North Pacific, Indian, and Southwest Pacific Oceans.[26]

All this extreme weather begs the simple question: Why? What is going on?

As with tornadoes, the combination of warmth and wetness shapes these storms. As our oceans capture heat from climate change, and the farther beneath the surface those waters warm, hurricanes become more powerful. If ocean waters turn chilly only feet beneath their surface, the winds of a hurricane churn up that cold water, and a storm naturally breaks itself. But that happens less often as ocean waters warm and spread that warmth deeper and deeper. That also helps explain why hurricanes stay stronger longer after landfall, fueling extended journeys further inland. A November 2020 *Nature* study, examining seventy-one Atlantic hurricanes with landfalls since 1967, found storms holding their strength almost twice as long. Hurricanes declined to two-thirds of their initial wind strength within seventeen hours of landfall in the 1960s, but storms now require thirty-three hours to weaken that much. "This is a huge increase," states study author Pinaki Chakraborty, professor of fluid dynamics at Okinawa Institute of Science and Technology in Japan. "There's been a huge slowdown in the decay of hurricanes."[27]

As our world warms, then, hurricanes have become more than an increasing threat to our coastlines. Inland cities will also see more damage from these increasingly persistent storms. Exactly how this plays out remains uncertain. As with tornadoes, complicating factors exist. Increases in wind shear tied to climate change may mean fewer Atlantic tropical storms, but those that do form will tend to be stronger and wetter, as was Hurricane Harvey when it swamped Texas, and Houston in particular, in August 2017.[28] And by the way, this is not only a concern for Florida, the most hurricane-hit state in the union, and its neighbors along the Gulf of Mexico. Those states find themselves targeted 15 percent more often than all other U.S. states combined. But Hurricane Sandy, for example, greatly impacted New York and New Jersey in October 2012. Only a Category 1 when it made landfall, Sandy drove water surge roughly fourteen feet at Battery Park, flooding large parts of Lower Manhattan and more than 88,000 buildings. All told, it swamped fifty-one square miles within New York City, killed forty-four people, and caused $19 billion in damages.

Storms are brewing across this land, then, from abnormal heat, drought, and flooding to tornadoes and hurricanes. All this extreme weather shares the fingerprints of global warming.[29] And each individual event will compound the damage of the next, with less and less time for recovery in between. "Viewing weather events as independent occurrences is like trying to understand a movie by looking at a series of brief clips," warns Michael Oppenheimer, professor of geosciences and international affairs at Princeton University. "They are important plot points, but not the whole story."[30] Take Hurricane Maria as an example. It made landfall in Puerto Rico twenty-six days after Hurricane Harvey, more than two thousand miles away from Texas. But these two events are inextricably connected. Damage in Puerto Rico was magnified by an exhausted U.S. Federal Emergency Management Agency, stretched thin in financial and personnel resources from two previous storms, especially Harvey. This is the future we face because of climate change. More extreme weather is in our forecast.

5

THE MELT IS ON

My mother passed away a little over a month before this trip. Like many who lose a loved one, my sorrow still surfaces across multiple layers. Her passion for gardening and the boundless energy she poured into each new cultivation project are one example. When my youngest child eagerly prepares to plant his own little plot, I smile wistfully, seeing her enthusiasm in him. But what starts as a happy memory is tempered with a dose of sadness. I mourn memories not made. I wish for more time with her. And I wish for more time for her with my children. Loss teaches us many lessons, but perhaps none more important than the intrinsic value of time.

That's why I'm here in our forty-ninth state. I'm extremely fortunate to share this special experience with my family, but my professional objective is to better understand our disappearing cryosphere. Alaska is a window into how climate change affects the ice on our planet—and why that matters. We all would be wise to better grasp why this melt is on.

Mere balance is the more pedestrian loss I'm preoccupied with now, namely not losing it. We've been kayaking over an hour this crisp and cool Sunday morning, carefully hugging the picturesque, rocky shoreline from our port in Whittier, before taking a quick hot chocolate and granola bar break. While the water here is deep, at about 120 feet, the relative proximity of land eases my mind. That psychological advantage is lost, though, as we paddle across Passage Canal on the east coast of the Kenai Peninsula. It's about two miles wide here, but more statistically intimidating, the average depth is 1,128 feet, maxing out at 1,290. Admittedly, according to nautical charts, the 767 feet of water at our crossing midpoint is well below both those numbers. But that's not exactly reassuring to sea kayak novices like us, no matter what my two older, teenage children may assert.

Passage Canal is a bay of Prince William Sound. Yes, the same Prince William Sound where the *Exxon Valdez* infamously breached its hull and spilled ten million gallons of crude oil in March 1989. You may have seen pictures of that tragedy and attempts to clean it up. Those indelible images, with oil washing ashore and an ecosystem devastated, first brought my attention to Alaska decades ago.

But the story of this land goes back much further. Native peoples called it home for at least ten thousand years before the Russians claimed it in 1741. Alaska fell into American hands over a century later when U.S. Secretary of State William H. Seward negotiated its purchase for $7.2 million. Many of us were taught that this acquisition was publicly scorned, often referenced as Seward's folly or Seward's icebox. A more careful analysis, though, shows this great land, a truly appropriate derivative translation from the native Aleut language, was long rumored rich in resources. Our last frontier, Alaska was recognized as a valued asset from early on. And at a cost of only about two cents an acre, the U.S. Senate voted to approve the treaty by a notable count of 37 to 2.

That didn't bode well for the original inhabitants—or the land itself. The 1867 Treaty of Cession with Russia denied "uncivilized native tribes" their rights as citizens, and their land became a larder waiting to be raided, our "last porkchop" if you will, as the eloquent environmental essayist Edward Abbey once described it. After seventeen years of rule by the U.S. Army, Treasury, and Navy, respectively, the Organic Act of 1884 and a Second Organic Act in 1912 created official governing structures, the latter formally establishing Alaska as a U.S. territory with an elected legislature. But it was World War II that shaped the Alaska we know today. "[That] buildup of population, transportation networks, and concomitant infrastructure is still felt," as historian Walter Borneman writes.[1]

Alaskan history is also defined by exploitation of its natural wealth. Three "gold" rushes top that list.[2] The "soft gold rush" for fur in the mid-eighteenth century first attracted Europe's powerful nation-states. A "hard gold rush" followed years later in the 1890s, with the *Excelsior*'s July 1897 landing in San Francisco spurring the Klondike stampede for gold. And finally, the "liquid gold rush" for oil began with construction of the Trans-Alaska Pipeline in the 1970s.[3]

Prudhoe Bay crude first flowed through the Tans-Alaska Pipeline in 1977. It's a marvel of engineering. Towards the end of our Alaskan adventure, after some interviews at the University of Alaska, I convince my family to see a stretch of this four-foot-wide, eight-hundred-mile-long pipe north of Fairbanks. Our kids weren't too keen on this part of the itinerary, but the promise to pan for gold at a tourist operation across the road swayed them. More than half the pipeline runs above ground. This prevents the hot oil from melting permafrost as it passes through the pipeline. Something called a VSM (vertical support member) allows the pipeline to ride saddle-like atop a metal crossbeam where it also slides back and forth as the pipeline expands and contracts with changing temperatures. This construction also creates crucial flexibility if an earthquake occurs.

That's not a hypothetical, by the way. Four out of five earthquakes in the United States occur in Alaska, an astounding 43,000 annually. We couldn't use the YMCA gym in Anchorage two weeks earlier because of structural damage from one the year before. Impacts from the largest quake ever recorded in U.S. history, the 9.2 magnitude 1964 Good Friday earthquake, are still visible from Anchorage to Whittier to Seward as we crisscross parts of the state. That said, the real reason I wanted to make this detour to the pipeline, aside from personal curiosity, was to show my children exactly how we receive this black gold, and to better understand how Alaska's fortunes rise and fall with our oil economy. Thanks to oil revenues, the state has no income tax and no statewide sales tax. While fishing, mining, and tourism account for most jobs, the state budget depends almost entirely on the oil and gas industry, whose taxes cover 90 percent of Alaskan revenues. There is even something called the Permanent Fund Dividend, an annual payout to every man, woman, and child since 1982. These typically are above $1,000 per person but at times exceed $2,000.

This means Alaskans view oil prices very differently from the rest of Americans. While most of the country celebrates when oil prices fall, Alaska does not. The state shares more in common with countries like Venezuela than states like California in that respect. That's painfully obvious as I make my interview rounds at the University of Alaska in Fairbanks. The governor recently proposed a 41 percent cut in state funding for higher education in FY 2020, amounting to $130 million lost revenue for the university system. It's a political move, of course, but also one rooted in a more than billion-dollar deficit as oil prices dropped from $85 to $55 per barrel that past year.[4] A compromise was finally reached a few weeks after I depart, but the university system still lost twenty-five million in 2020, twenty-five million more in 2021, and an additional twenty million in 2022.

Alaska is in a tough spot. Because of its size, varied topography, and climate, the state spends per capita almost three times the national average on its state budget. Unfortunately, this means the standard solution to budget problems is often drill for more oil. As such, Alaska is an exaggerated microcosm of our society's dependence on fossil fuels. Alaska is also, like my home state of Florida, overly dependent on natural resources more generally, namely the tourism related to those resources. Both corners of our country would benefit from diversification.

As we have seen in previous chapters, several locales stand out when we think of climate change impacts. Add Alaska to that list. For one, it's a lot hotter here than it used to be. The day before we arrived, Anchorage registered 90 degrees Fahrenheit, breaking the previous record by an eyebrow-wiping five degrees. Other parts of the state see temperatures twenty to thirty degrees hotter than normal throughout that July, making it the hottest month in Alaskan recorded history. Thanks to climate change, Alaska is our fastest warming state, with temperatures rising twice as fast as our other forty-nine. That's true throughout the Arctic. Its roughly two million square miles, divided among the eight Arctic countries of Canada, Denmark (Greenland and Faroe Islands), Finland, Iceland, Norway, Sweden, Russia, and United States, are warming twice as fast as the rest of the planet.

Fueled by abnormally high temperatures and correspondingly higher evaporation rates, multiple wildfires burn across Alaska throughout our stay. We smell their smoke from time to time, including within our hotel hallway last night. But the air is fresh and free of smoke today as we admire several glaciers from atop the Chugach and Kenai Mountain ranges surrounding Passage

Canal. And seeing these with your own eyes is a powerful reminder that these rivers of ice have undergone continuous retreat over the last four decades due to climate change.[5]

The small town of Whittier is near the head of the bay and the only settlement along Passage Canal. It is also known as the Weirdest Town in Alaska.[6] That's a badge of honor in these parts—and one won among some healthy competition. In about a week, for example, we will explore Talkeetna, a village that reportedly inspired the quirky 1990s television series *Northern Exposure*. Fairbanks as well as Chena Hot Springs, just to mention a few more, also have their defenders of the strange and unusual. But Whittier, and its decided mix of contradictions, gets my vote. Gateway to Prince William Sound, even though it's difficult to reach, Whittier is a deep-water port that remains freeze-free throughout the winter. This fact, combined with a semipermanent cloud above thanks to those nearby converging Chugach and Kenai Mountain ranges, made Whittier an ideal location to hide our far-north Pacific military base from Japanese bombers during World War II. Afterward, with the ensuing Cold War, the military stayed another fifteen years.

Aside from sea access, the only route in and out town is the 2.5-mile Anton Anderson Memorial Tunnel. Sixteen feet wide, it accommodates only a single rail line. That was adapted twenty years ago to share access with one-way auto traffic once an hour, but it still closes from 11 p.m. to 5:30 a.m. each day, and even longer in winter, making last openings a little stressful. As writer Mark Adams describes, Whittier is essentially sequestered "from the rest of Alaska like a geographic safe room."[7]

With a population of roughly 208, the town is home to more boats than people. And interestingly, most of those 208 individuals live under the same roof. It's a big roof. The 15-story Begich Towers, with an exterior, as Adams explains, that "could pass for one of nicest buildings in Pyongyang," is self-contained and houses a 196-unit condominium community, including a laundromat, police department, post office, and grocery store.[8] Moving a little further up the road is the Buckner Building, an even larger complex, but one eerily abandoned since the Army departed in 1960. Alaskan weather has essentially served as its absent-minded caretaker for the half century plus since, with the 1964 Good Friday earthquake adding some structural damage for good measure. Some say it's the perfect postapocalyptic movie set.

That dreary description resonates well while patrolling its perimeter. Built as housing for enlisted military personnel, with a 350-seat movie theater, bowling alley, shooting range, bakery, and even a jail, it was once the largest building in our largest state. Now, it calls to mind images of Ukraine's Pripyat, the desolate town once housing Chernobyl workers until the nuclear meltdown in April 1986. Each now provides shelter for wildlife as well as an attractive canvas for graffiti artists and the occasional thrill-seeker. One American urban skiing pair even filmed their crafty maneuvers a few years back, navigating the Buckner Building's stairwells, doors, and windows in what they label online as "Five Floors of Fury."

We're back below that abandoned structure now, receiving a safety lesson on our fiberglass double kayak from the staff at Prince William Sound Kayak Center. It rains often in Whittier, almost daily, with the sun shining only a third of the year. But this morning we are lucky. While we're drenched later during an afternoon hike above town, the clouds show mercy during our three-hour morning paddle. Stunning sea cliffs and a lovely waterfall near a bird rookery sheltering ten thousand black-legged kittiwakes would be the highlights—if not for the glaciers tucked amidst the mountain passes. Turns out we were even more fortunate in our timing. The next morning, I open the *Anchorage Daily News* to learn a tragedy struck Whittier only hours after we departed. Just a stone's throw from where we put in our kayaks, a deadly explosion sank a ninety-nine-foot salmon fishing vessel, and the ensuing fire spread to destroy a forty-foot section of the city pier.[9] A man my age lost his life.[10] Alas, Alaskans are all too familiar with loss. And climate change now brings loss of an altogether different magnitude.

Entire villages will disappear.

Take Shishmaref as an example. Just north of the Bering Strait, it's on the small barrier island Sarichef, which has lost nearly three thousand feet since 1980. The island is accessible only by airplane or boat, unless winter ice is thick enough for snowmobiles. About a quarter-mile wide at its center, it sits strategically between the Chukchi Sea and the Serpentine River estuary, which makes it prime grounds for hunting and fishing. Islands like this used to be protected for most of the year as sea ice served as a natural sea wall. Now, due to warmer temperatures, ice forms later. This means less protection as fall and winter storms increasingly erode its coast. This Alaskan village is slipping into the sea.

"Decades ago, ice was fully formed by October. Now it takes to December, and ice isn't thick enough to cross it safely," explains Esau Sinnok, a Shishmaref native and Arctic youth ambassador in a program started by the U.S. Interior and State Departments.[11] The island made national news in August 2016 when its largely Inupiat tribe narrowly voted 89 to 78 to relocate their ancestral home. This won't be cheap. "About fifteen years ago, they estimated the cost at $180 million, but I would figure it's much higher now," Donna Barr, secretary of the Shishmaref Council said. "We don't see the move happening in our lifetime because of the funding."[12]

Shishmaref is not alone. Over thirty other Alaskan villages face similar threats, according to a study by the U.S. Government Accountability Office. Decline of sea ice combines with more intense storms to drive much of this. Rick Thoman knows this all too well. Thoman is a climate specialist at the University of Alaska, but he calls himself a climate translator. With a foot in multiple realms, Thoman talks modeling and high-end statistical analysis with climate scientists, then translates that information for people like you and me who aren't specialists in the field. In his well-practiced radio voice, honed over a thirty-year career with the National Weather Service, Thomas explains, "In western and northern Alaska, along the coast, by far the single most important thing in winter is not temperature. It's not precipitation. It's sea ice."[13]

John Walsh echoes that same concern from his International Arctic Research Center office, as birthday party chatter seeps in from a neighboring lounge late one afternoon at the University of Alaska Fairbanks campus. Walsh is an expert in Arctic climate, weather, climate change adaptation, and sea ice. He tells me about a report he was working on with Thoman called *Alaska's Changing Environment*. The Alaska Center for Climate Assessment and Policy and the International Arctic Research Center both helped shape the report, which describes major changes in temperature, sea ice, glaciers, permafrost, plants, animals, and oceans. As it states, "The extent, duration, and thickness of sea ice has changed significantly in the seas around Alaska. The changes have been most widespread in the late summer and autumn. The average sea ice concentration . . . shows much lower ice concentrations (or no ice) in the Chukchi and Beaufort Seas in 2018 compared to thirty years prior."[14] In fact, annual September minimum sea ice extent within the Arctic Ocean has decreased at the rate of 11 to 16 percent per decade since the early 1980s, with accelerating ice loss since 2000.[15]

Since this is ice already in the ocean, like ice in your water glass, it does not raise sea level. Only ice melting from land would do this. But when sea ice melts, it still affects temperatures as we lose an important cooling function, one akin to turning off an air conditioner. It also means less protections from stronger storms for islands like Sarichef. And there is the albedo effect. People often wear white on hot days to reflect the sun and keep a little cooler. When incoming sunlight strikes ice in the Arctic Ocean, as much as 70 percent reflects into space. Not so with open water. As sea ice melts, much darker open water takes its place, reflecting only six percent of the sunlight. The remaining 94 percent is converted into heat. The same principle applies to ice that only partially melts, whether on land or water. When snow melts, particles within it become trapped on the ice's surface. Some of these particles melt away with meltwater, but the more the snow melts, the more particles build up on the surface, further darkening any remaining ice.

This partly explains why temperatures in the Arctic are rising at twice the speed of the rest of our planet. The pole is changing from a heat reflector to heat absorber. As research professor at Columbia University's Lamont-Doherty Earth Observatory Marco Tedesco writes with his coauthor, reporter Alberto Flores D'Arcais, "What used to be Earth's refrigerator has become a huge sponge leaking water into the ocean."[16] Scientists like Tedesco borrow from pop culture to best explain the frightening implications in Greenland. Here, glimmering gemstones of melt water rest amid the ice sheet. But don't let their beauty deceive you. These are what scientists call cannibal lakes. Feeding on themselves, they slowly gouge away more ice, creating more melt.[17]

The truth is that glaciers like those in Greenland and Alaska are melting across the globe, from the Andes and Alps to the Rockies to the Himalayas and Tibetan Plateau. By 2050, aside from Greenland and Antarctica, most will be gone. Across the Arctic, melting season still begins around early June, but the refreezing date comes later each year. In the 1970s, the melting season ended in September. Now it's not until October. These remnants of the Cordilleran Ice Sheet, from tens of thousands of years ago, when much of the United States was covered in ice, are retreating not just because warmer temperatures melt them. They also are retreating because warmer temperatures mean more precipitation falls as rain rather than snow, and that rain further accelerates snow melt. "More than 90 percent of Alaska's glaciers are retreating," according to Thoman and Walsh. "Between 2002 and 2017, Alaska glaciers thinned on average by several feet per year. Overall mass loss during this period was nearly 60 billion tons of ice per year."[18]

Of course, as you'll recall from chapter 1, if every alpine glacier across the globe melted, sea level would only rise about a foot or two. The real threat in terms of sea level rise comes from ice sheet melt in Greenland and Antarctica combined with thermal expansion.[19] Particularly worrisome, as also noted in chapter 1, is West Antarctica. Its marine-terminating glaciers are known as tidewater glaciers. Massive chunks lie below sea level, so, when their floating ice shelves break off, sea level does not rise. But these ice shelves do buttress glaciers behind them. As they slide from land into the ocean, sea levels will rise—significantly.

And critically, this melting will not be felt equally around the world. Melting Greenland is already having a bigger impact in southern latitudes, paradoxically, while Antarctic melt is felt more in northern realms. As these ice sheets become smaller, their gravitational pull upon the water around them lessens, so sea levels in the immediate area fall – but that relocating water pushes water higher on the opposite side of the planet. Exactly how fast this melt unfolds is difficult to predict. One major recent melt event on the Antarctic peninsula was the 2,200-square-mile Larsen C, the size of Delaware.[20] It began to crack visibly in 2010 before breaking off in 2017.[21] Small changes in the jet stream or cloud cover, mirroring chaos theory, could bring more breakoffs and quicken future melt.

Alpine glacial melt, like that here in Alaska, also deserves our attention. It won't be nearly the factor in sea level rise as the melt from Greenland and Antarctica we've noted, but its loss fundamentally shapes precipitation patterns. This brings two very different complications. Without glaciers, water is no longer available in the form of traditional summer glacier runoff. Drought results. And when rain finally does come, flooding, erosion, and crop damage are even more likely. Impacts extend far beyond where glaciers flow. In Central Asia, for example, roughly half a billion people depend upon meltwater from the Tibetan Plateau. Here in the United States, two-thirds of the Cascades region's electricity comes from hydropower sourced by glacial melt.

Permafrost, the soil, rock, or ice that remains frozen over two years, is the centerpiece to Alaska's melt. As Alaska warms, its soil is getting warmer, and the depth of seasonal thaw, known as active layer depth, is increasing. On the Seward Peninsula, for example, measurements show thaw depth reached thirty-three inches in 2018, compared to twenty to twenty-four inches earlier that decade.[22] This wreaks havoc on infrastructure. As ice melts and loses volume, land slumps, and the stuff we build on it damaged. "Permafrost is really important in Alaska," elaborates Larry Hinzman, vice chancellor of research and professor of civil and environmental engineering at the University of Alaska Fairbanks. "The presence or absence of permafrost really affects everything. What do we do to adapt? It's not going to be inexpensive. It costs a million dollars a mile to repair rural roads."[23]

But more than infrastructure and tipsy trees are at stake. "It's a very noisy, dynamic topic," Hinzman adds. "The effect of permafrost thaw in the Arctic will affect sea level rise. It's a complex, interrelated system. Change in one component of the system reverberates. It's all interrelated."[24] "In Alaska, the primary control of water is through permafrost," continues Hinzman, whose primary research interests involve permafrost hydrology. "Where you have permafrost, the ground is quite wet. But in Alaska we are actually quite dry. We don't get much precipitation. Where you don't have permafrost, the soil is quite dry. The amount of water in soil really controls what kind of vegetation grows there. It controls what is susceptible to fire. Water really controls everything, but permafrost controls water. You only need permafrost to thaw one meter from the surface to cause dramatic change if it doesn't freeze back."

Permafrost starts a few feet below the surface and extends as much as hundreds of feet down. At the North Slope, some permafrost drops over 2,100 feet below the surface, according to Vladimir E. Romanovsky, another permafrost researcher at the University of Alaska Fairbanks. Notably, this permafrost contains vast amounts of carbon in its organic matter, plants that consumed atmospheric carbon dioxide centuries ago, then died and froze before decomposing. "Permafrost contains twice as much carbon as is currently in Earth's atmosphere," according to Thomas Douglas, a geochemist in the U.S. Army.[25] And as this organic matter thaws, microbes convert portions to the greenhouse gases carbon dioxide and methane. Scientists predict this could raise temperatures another 1.7° F.[26] The entire process may take centuries, perhaps longer, but within a couple of decades, permafrost melt may overtake China as the leading source of greenhouse gases.

Not all of this is in Alaska—but some is. And that means the state is shifting from a net sink to a net source of carbon.[27] "We have evidence that Alaska has changed from being a net absorber of carbon dioxide out of the atmosphere to a net exporter of the gas back to the atmosphere," contends Charles Miller, a chemist at NASA's Jet Propulsion Laboratory who measures gas emissions from Arctic permafrost.[28] That's a big deal. Alaska's Tongass National Forest is larger than the state of West Virginia. Its Juneau Icefield alone is larger than Rhode Island. All total, Alaska can hold our second (Texas), third (California), and fourth (Montana) largest states—with space remaining to hold Hawaii, New England, and a handful of cities. When something this size starts changing dramatically, we best take notice. And seeing climate impacts like this firsthand underscores why people from a range of political ideologies, from conservative to liberal, should share an interest in halting that change.

"What Aldo Leopold calls the land ethic, we have not really built that into our society," points out Rose Keller within her office in Denali National Park's headquarters.[29] That's a mistake. Alaska's melt is evidence to her point—that we must incorporate a land ethic into our daily lives soon. We need to better understand, like Leopold, that how we treat this land profoundly impacts our future.

Nothing illustrates this better than Denali's poop problem, sixty-six tons of it to be exact. That's the amount of frozen feces climbers have left on the summit, according to the National Park Service. And as the glacier melts, this excrement is unfreezing and flowing with meltwater to lower elevations.[30] National Park Service glaciologist Michael Loso began studying this a decade ago. "The waste will emerge at the surface not very different from when it was buried," he asserts. "It will be smushed and have been frozen and be really wet. It will be biologically active, so the *E. coli* that was in the waste when it was buried will be alive and well. We expect it to still smell bad and look bad." I wonder, is this inglorious image, this repulsive smell, enough to finally bring us to our senses when it comes to the threat of climate change?

6

CHANGING HABITATS AND SPECIES DIVERSITY LOSS

The sweaty but festive crowd, abuzz with excitement, gathers in tightly around the stage and elevated track. Lobstah Mobstah just clobbered the competition, sprinting, ok crawling, the roughly four-foot-long course in well under a minute. Unfortunately, my dollar bet was foolishly placed on Lance Clawstrong, who never even had the opportunity to be disqualified, content to rest his laurels at the starting line throughout the race. Next up for this Bar Harbor Fourth of July tradition is perennial crowd favorite, Suppah, and my son Malachi volunteers to be a guest starter, gingerly holding two of Suppah's crustacean competitors in their "starting gates" for this quirky contest.

National Geographic labels Mount Desert Island's Fourth of July festivities among the top ten in the land, and the *Today Show* once named this line-up of events our nation's best Independence Day celebration. We powered up appropriately this morning with a Maine-inspired blueberry pancake breakfast on the large athletic field at the edge of town, followed by a patriotic Independence Day Parade up Main Street. And, of course, there will be traditional fireworks along Frenchman Bay later this evening. But what really piques our curiosity are these annual lobster races benefiting the Mount Desert Island YMCA scholarship fund. Nowadays, when you think American lobster, you think Maine. Yet, ironically, by the time I retire, perhaps as early as 2037, American lobsters like those competing today will no longer reside in U.S. waters. This quirky national independence celebration will depend entirely upon Canadian crustaceans. The reason rests with climate change.

Maine has a long tradition of overfishing. Herring, hake, and halibut were all overexploited. So were croaker and cod. But not lobster. Protection of this lucrative resource dates to conservation measures in the 1890s. Nearly a century ago, in 1930, the state implemented a maximum harvest size. And in 1995, the state legislature established the seven rectangular management zones enforced today, running from Kittery in the southwest corner of the state to Cutler in the northeast. Initially, all that seemed to have paid off. The London-based Marine Stewardship Council even certified Maine lobster as sustainable in 2013.

Alas, climate change has something else in mind.

Maine's take within the U.S. lobstering industry was a healthy 50 percent back in the 1980s. Now, as southern New England stocks disappear, it's ballooned to 85 percent.[1] Maine's fisheries produced approximately 20 million pounds of lobster per year in the 1980s. Over the last decade that number has grown to a staggering 120 million pounds per year.[2] That's a welcome boon, one Maine profits from. But it's a boon that will not last. "The center point where stocks are most dense shifts 4.3 miles northeast each year in the open Atlantic," explains science writer and naturalist Christopher White.[3] No one knows exactly when this started, but it's been measured as far back as the 1960s.

Lobsters don't migrate in the classic sense. Each year their larvae simply reestablish farther north and northeast. One study by a Princeton University team tracking movements of 360 marine species over forty years, from 1968 to 2008, is telling. They found, as ocean temperatures climbed locally over two degrees Fahrenheit, that the American lobster migrated forty-three miles per decade off the New England coast.[4] Some locals challenge this trend, claiming Maine's lobster bounty derives from fewer predators as populations of groundfish like cod, which eat young lobsters, have crashed.

But looking closely at the data, the role of climate change is clear. Marine species have optimal thermal zones, with lobsters preferring chilly waters of 53 to 64 degrees Fahrenheit. And that cold zone they crave is moving north. This matters to New Englanders whose livelihoods depend upon them. American lobster is the most valuable fishery resource in North America. And in Maine, it's not just the 3,800 individuals holding lobstering licenses that have a vested interest. Another 10,000 people in the state provide support for the industry as pickers, cooks, and drivers. Even teenagers get in on the action, earning as much as $5,000 a week lobstering in the summer and bringing home a handsome $60,000 before returning to school in the fall.[5] All total, the Maine lobster industry produces over $1.7 billion annually.[6]

The Gulf of Maine is the size of Lake Superior and encompasses one-fifth of our Atlantic seaboard. Home to the greatest tides in world, it includes the rocky coastlines of Maine and New Hampshire as well as the northern half of Massachusetts. The basin is bounded to the east and south by underwater banks. Georges Bank on the southern side isolates the Gulf of Maine from the warm Gulf Stream, while the northern Labrador Current cools its waters. That makes it ideal lobster habitat, producing record sizes up to three feet long.

Yet, that ideal habitat is changing.

Ocean water temperatures, thanks to climate change, are rising across the globe, but the Gulf of Maine's waters are heating up faster than virtually everywhere else. It's already too hot for the aforementioned cod, historically the second most important fishery in the basin. "The Gulf of Maine, temperature-wise, is changing very rapidly compared to the rest of the world," says Dave Carlon, a marine scientist at Bowdoin College. "It's a little microcosm for what's going to happen in other parts of the world as the temperature changes."[7] Scientists believe the additional increase in the Gulf of Maine is tied to weakening of the Atlantic overturning circulation. That's preventing some warm water from reaching its traditional higher latitude destinations, causing the Gulf Stream to deposit these warm waters along the New England coast, instead.[8] While modest temperature elevations can increase metabolism and growth, temperatures above 68 degrees Fahrenheit stress lobsters, bringing shell disease, susceptibility to parasites, and a wasting illness that makes them unsellable, if it doesn't kill them first.

That leaves three options for American lobsters: flee to deeper waters, set up home further north, or perish before reaching our plate.[9] Actually, a fourth option to survive within Maine waters exists as well, but only if we mitigate climate change. "If the Paris [climate change] goals are met, Maine holds on to a lobster fishery," explains Andrew Pershing, a scientist at the Gulf of Maine Research Institute. If not, *Homarus americanus* may require a new scientific name in perhaps a little over a decade.

These thoughts are far from my mind now, though. A wet morning chill still holds its grip within this boreal forest of spruce and fir, even as streaks of warm sunlight pour through the coniferous branches high above. Patches of moss and lichen, alternating between shades of green and white, carpet the rustically decorated forest floor, as do countless fallen branches, decaying tree trunks, and the occasional protruding boulder. Such beautiful walks in the woods have drawn summer crowds to Acadia National Park for over a century, but we are off that beaten path, on a much smaller island across from the main portion of the park nestled within Mt. Desert Island. Randomly placed boulders and branches necessitate frequent adjustments to reach our GPS-directed objective a little under half a mile away.

These periodic pauses provide a chance to reflect on our surrounding beauty. They also offer a glimpse into how this island sounds. Yes, sounds. It's the week before the Fourth of July rendezvous with my family in Bar Harbor, and I'm here as a volunteer citizen scientist with the environmental group Earthwatch. You could say my seven colleagues and I are marching to our own beat, helping a forest ecologist gather data for his study here in Acadia. Or you could say, as Earthwatchers, we merely march to the earth's beat. Mother Earth is our drummer.

Either way, that beat is changing.

Forest ecologists like Nick Fisichelli document shifts due to climate change with the introduction of new species and decline of previously established ones. His work shows this boreal island forest already looks remarkably different from the first time a group of Earthwatch volunteers gathered data here a quarter century ago. But sight is not the only sense that is changing. Climate change is also altering how this island sounds. That might sound a little odd at first. But think about it. The sounds of nature can be as familiar as crickets chirping you to sleep at home—or as mysterious as the late-night call of a coyote rousing you awake far afield. The point is these sounds have context, natural context. Forests, deserts, rivers, and oceans all hold distinctive sounds, sounds we learn to associate with geographic locales.

The floods, droughts, wildfires, hurricanes, and tornadoes discussed earlier will change how these sounds play out. Loss of biodiversity spurred by climate change will as well. Some notes will be immediate and obvious, almost staccato-like in their rhythm. Others, at least initially, resemble an adagio composition, softly and slowly building to a climatic closing as habitat ranges shift and ecosystems evolve. "Shifting weather patterns are already known to affect weather-dependent phenomenon like bird migration and wildflower blooms, ultimately cascading through populations and ecosystems—but the role of sound in all this has received relatively little attention," assert ecologists Jérôme Sueur of France's Sorbonne University, Almo Farina of Italy's Urbina University, and independent scholar Bernie Krause.[10]

Leaders in the field of soundscape ecology, a discipline created in the late 1990s, these researchers examine the full spectrum of biologically produced sounds in a habitat, from singing birds to chirping crickets to rustling leaves. While previous research on biological sounds focused on single species, their work on "biophony" concentrates on interactions between these species. For example, a male frog's mating success is shaped by more than its ribbits. Vegetation changes may alter transmission of that sound, striking a chord of ramifications for not only the frog's progeny but countless other species interconnected within their food web. Sueur, Farina, and Krause fear this may reduce species diversity, homogenizing surrounding habitat.

That's not good. Diversity adds resilience to nature's portfolio. Without it, akin to undiversified investment accounts, we are left vulnerable to the ups and downs of our changing climate. It's a lesson we should know better by now. But we don't. My Earthwatch colleagues and I are helping gather data that we hope will correct that.

We had to work to get here, climbing innumerable boulders along the shoreline to reach a sharply angled hillside stretching a dozen or so feet into a patch of low-lying prickly bushes before reaching the forest edge. But this adventure is not as rigorous as previous stints I've had with Earthwatch, including sixteen days tracking large wildlife in Africa's oldest park, South Africa's Hluhluwe-iMfolozi, and enduring 4:30 a.m. wakeup calls to record macaw nesting activity in the Peruvian Amazon. This time around, rather than tenting within electrified fences to deter culinary curious lions or fending off Amazonian sandflies that carry the parasitic disease *Leishmaniasis*, I'm enjoying the comforts of a civilized mattress and warm shower. There's even a treadmill and television in the common laundry building next door. That said, this trip still offers plenty of adventures, all rooted in experiential learning and gorgeous scenery.

There're some inspiring fellow Earthwatch volunteers as well, including two quintessential role models. These two ladies are octogenarians, passing that life landmark nearly a half decade ago but showing little signs of slowing down. To the contrary, they continue to push themselves daily, constantly improving their understanding of our world as individuals and helping make the planet a better place for the rest of us collectively.

A similar type of thinking sparked creation of our national park system, saving special places across this land for public use and appreciation. President Woodrow Wilson first granted federal status to the land now known as Acadia in July 1916. Back then it was called Sieur de Monts National Monument, renamed as Lafayette National Park less than three years later.[11] That designation made it our first national park east of the Mississippi River. Its location here in New England makes it a popular destination, and park attendance set a record of 4.07 million in 2021, with most of those visiting in the summer months. Acadia encompasses more than 47,000 acres, including half of Mount Desert Island, part of Isle au Haut, the tip of the Schoodic Peninsula, and sixteen smaller outlying islands like this one we are studying today. Well established trails and an historic carriage road system financed by none other than John D. Rockefeller crisscross the popular Mount Desert Island portion, which is capped by Cadillac Mountain, at 1,529 feet the highest mountain on our eastern seaboard.

This park is evidence that one individual, to borrow from legendary cultural anthropologist Margaret Mead, really can make a difference. Known as the Father of Acadia, George B. Dorr devoted the better part of his life to first establishing and then expanding Acadia. It helped that he had spare time as a bachelor—along with substantial inherited wealth. One of the original "rusticators," Dorr received a letter in 1901 from Harvard President Charles W. Eliot requesting assistance in forming a committee to "hold reservations at points of interest on this Island, for the perpetual use of the public."[12] That work became Dorr's obsession for the next forty-seven years as he grew the park from mere concept to more than forty thousand acres.

Heading back to our week-long base at the Schoodic Education and Research Center, about an hour's drive around Frenchman Bay from the port in Bar Harbor, I remind myself great deeds like Dorr's often require a healthy dose of perseverance. The Schoodic Education and Research Center is a testament to that and has become the largest of nineteen research learning centers in the national park system. It's truly a living laboratory. "We are science storytellers," explains Abe Miller-Rushing, a phenologist (one who studies the timing of recurring natural phenomena in relation to climate) and primary investigator for this Earthwatch grant, during his presentation to our group early that evening. "You with Earthwatch are part of changing our culture in the NPS to increase use of science within parks," he emphasized.[13]

Science, including citizen science conducted by amateurs like me, was instrumental in Acadia National Park's creation. It's written into the founding legislation. Acquired by Acadia National Park in 1929, Schoodic Point is the headquarters of that operation today. Schoodic frames the entrance to Frenchman Bay upon its eastern edge and is decidedly more secluded than the main body of the park on Mount Desert Island. Only about 10 percent of park visitors ever see it.[14] The U.S. Navy was the main influence for much of the twentieth century, relocating a strategic listening station here from across the bay in 1935 and expanding operations during the Cold War, before closing in 2002 and transferring the land to park scientists.

We are here to assist with a variety of studies, including construction of a passive heat experiment as well as recording phenology test plot data. Where we get our best taste of scientific fieldwork, though, is documenting vegetation change and balsam fir, eastern hemlock, red pine, red spruce, and white spruce tree growth around the park.[15] Our task is to track down old rebar marking ten-meter square plots randomly created with GPS coordinates years ago, set new rebar rods marking the northwest corner of those plots, and conduct a tree survey as well as shrub cover estimate of them. This is what is called citizen science.

After another stellar dinner, featuring the official Maine state treat of whoopie pie, I sit down with Nick Fisichelli, the Forest Ecology Program Director for the Schoodic Institute at Acadia National Park, to explore his use of volunteers like me. As kitchen cleanup sounds create a din in the background, Fisichelli mentions he had the pleasure of working with Earthwatch volunteers over two decades ago while volunteering himself at

Oregon Caves National Monument. "I still remember high quality, high caliber people, really interested in the science," he reminisces. Our group is his first this summer and only his second stint leading volunteers. Last year he supervised a dozen citizen scientists, including a teenager with their parent. This year he's exploring the other end of that age spectrum, one that might well hold an Earthwatch record. To Fisichelli, his work and the Earthwatch mission are a perfect match, both seeking to build interest in science and the use of it. "Rather than being some ivory tower producing results in a study that just collects dust somewhere, this is about getting people involved, understanding how science works," Fisichelli explains. "There's a huge benefit of citizen science. In some ways it is as valuable as the actual science we carry out."[16]

This education piece is central to Earthwatch's mission. And Fisichelli gets it. "This is about understanding our evolving role in the world," he continues. "I think of our work as a marriage, always changing. Citizen science with Earthwatch offers an opportunity to discover more about nature and our relationship to it. Whether lobster or clams, or timber or tourists, you need science to understand what is happening and to make better informed decisions."

It might not be obvious amid today's intermittent downpours mixed with constant drizzle, as the chilly June temperature hovers around our new global annual average of 59 degrees Fahrenheit, but Acadia National Park is an ideal spot to study climate change. Acadia is juxtaposed between deciduous forests to the south and boreal spruce-fir forests to the north as well as unique sea edge where cold northern ocean currents mix with warm and moist continental air.[17] These edge effects create vulnerabilities. Over the past 120 years, the park has lost 18 percent of its native plant species.[18] Within the last half century, five hundred new insect species have come to call Acadia home.[19] Bird migration has changed throughout the region, too.[20] "These changes may be creating mismatches among predators and prey, plants and pollinators, and consumers and resources by changing the timing of when various species need a resource and when that resource is available," write biologists Richard Primrock and our host, Abe Miller-Rushing.[21] Fruits ripen earlier, before migrating birds return to eat them, which translates into food source deficiencies when those birds do arrive. Since its 1916 founding, Acadia's growing season has lengthened nearly two months. Warming during winters and nighttime, in particular, threatens native plants while inviting invasive species such as black legged ticks that thrive in the longer, warmer summer season—and carry Lyme disease.

There is also more rain. Across the state, Maine's annual precipitation increased six inches per year from 1895 to 2013. Extreme rainfall events are more likely, too, occurring every dozen years now instead of every fifty. This damages infrastructure in Acadia. More than a thousand culverts, for example, were built in the 1930s with much different rain patterns in mind, too small to carry today's increased load.[22] Sea levels are rising, too. Here in Acadia it's seven inches higher than a century before. And across the globe, Acadia included, the ocean is acidifying more rapidly than any other time in the last 300 million years. Other Earthwatch groups will help study those oceanic impacts, but we're concentrating on the land this week, namely these emerging phenological mismatches and the loss in species diversity that accompanies them.

These changes are dramatic—but not always visible. You may have seen headlines a couple years ago warning that one million species are at risk due to climate change. Extinction is natural. But the pace today is decidedly not. It's one thousand times higher than the expected background extinction rate, according to legendary entomologist E. O. Wilson. Habitat loss is most responsible. Other factors include hunting, general pollution, and invasive species that outcompete native species for necessary resources. Climate change is still another. We now face the sixth mass extinction in our planet's history. Of the five previous, all but the one that killed the dinosaurs were shaped by climate change.

Roughly two hundred species of plants and animals become extinct each year.[23] If worldwide average temperature increases 3.6 degrees Fahrenheit, we may see a total of 400,000 species become extinct. Alas, the enormity of such species loss is difficult to grasp unless our level of analysis focuses upon local scale.[24] "Spatial context really matters," explains Rebecca Chaplin-Kramer, lead scientist at the Natural Capital Project, a Stanford University–based research group. "It's not just the total amount of nature we have, but where we have it, and if it's in the place where it can deliver the most benefits to people."[25] The U.S. National Phenology Network is attempting to better grasp such relationships, gathering leaf-out and bloom data along with information about when species migrate and reproduce. The network officially began in the mid-2000s, but recorded observations date back to the 1950s. That data shows several key trends. Trees flower earlier. Birds migrate to breeding grounds earlier or lay their eggs sooner. Butterflies relocate higher up mountains or to cooler regions closer to the poles.

And yet many people have not experienced these effects personally. Mark Hineline, a history of science professor at Michigan State University, wants to see that change, calling on nonscientists like you and me to practice phenology in their own backyards.[26] Like Earthwatchers, we can all observe how plants change throughout the year—without ever leaving our own yards. If the urge to explore farther afield strikes you, iconic examples range from the lovely pinkish-white Japanese cherry blossoms in our nation's capital to these enchanting forests of New England.[27] Take maple syrup, for example. Climate change will alter the production of that breakfast table nectar, according to a 2019 Dartmouth study. "As the climate gets warmer, the sugar maple tapping season will shrink and get closer to a December date," says David Lutz, a research assistant professor of environmental studies who examined six sugar maple stands from Virginia to Quebec. As Lutz explains, two climate-sensitive factors shape maple syrup production. Sugar content is determined by the previous year's carbohydrate stores while sap flow depends on the freeze-thaw cycle. Like the lobstering industry, there will be winners and losers when it comes to future maple syrup production. Canada's current 80 percent share of world production will increase, while U.S. states like Virginia and Indiana, by 2100, will effectively bow out of the market altogether. New Hampshire and Vermont will also lose substantially.[28]

Lobster season. Maple syrup season. These might not seem as threatening as more intense hurricanes or larger, more frequent wildfires addressed in earlier chapters, but such seasons are central to New England's sense of place. And they are important economic drivers, from farming and forestry to tourism. It's not just New England that will suffer when it comes to ecosystem shifts and biodiversity loss. In one study examining 1,460 different species of plants, birds, mammals, reptiles, and amphibians across the continental United States, 14 percent will not survive the current climate change underway. Others will be forced to relocate. The state symbol for Ohio will likely be spared from extinction, for example, but researchers believe it will no longer call the Buckeye state home. Committing college football sacrilege, the tree's territory will shift north into archrival Michigan.[29]

In another noted study, analyzing the records of the geographical ranges of more than 1,700 species of plants and animals, Camille Parmesan, a University of Texas at Austin ecologist, and Gary Yohe, an economist at Wesleyan University, found ranges moving, on average, about four miles per decade toward the poles. But many species simply won't be able to migrate quickly enough if our planet continues to warm. "This is not an abstract concept anymore," argues Evergreen State College professor John Withey in describing yet another study he coauthored.[30] "We need to take action as soon as possible, thinking about where species may need to go under climate change, and providing corridors through which they can move."[31]

Turning to an American treasure, the naturalist, essayist, poet, and philosopher Henry David Thoreau, provides further insight on this need for wildness. Thoreau famously declared, "In wildness is the preservation of the world."[32] He also offered insight on where to find that wildness. Thoreau preferred not to stray far from his Concord home, leaving only for short trips such as these forests in Maine. He even once boasted, "I have travelled the world . . . without leaving Concord." Thoreau's local adventures, and the powerful insights he gathered from them, prove there is much to learn from our immediate surroundings.

Similarly, we need not travel to the far corners of the earth to better understand global climate change. Its impacts can be found locally throughout our own country, throughout this land. Thoreau's idea of wildness, furthermore, is often misquoted, with admirers mistakenly substituting the word "wilderness," instead. While the error is understandable, it unfortunately conveys almost the opposite of Thoreau's intentions. Wildness refers not only to remote and awe-inspiring places but also to our immediate surroundings. As such, Thoreau exposed our problematic separation of people from nonhuman nature. Travel around this land, with an emphasis on destinations closer to our homes, follows Thoreau's logic. Like him, you need not go far. Wild nature is much closer than often recognized. Take time to see it.

Colonies of flying foxes famously reside within the sprawling botanical grounds of downtown Sydney, Australia. Birds of prey often take refuge in and hunt from metropolitan high rises across the United States. There's at least one racoon family that periodically ventures out for late night suburban snacks in my Central Florida neighborhood. This diversity bestows resilience, enhancing our entire ecosystem.

Yellowstone National Park rangers learned that lesson when reintroducing wolves in 1995 after an absence of nearly seventy years. Within the Lamar Valley, for example, elk were decimating aspen stands, feasting upon young aspen shoots before they had a chance to grow. But wolves eat elk, an average of one elk per wolf every month during the winter. So, when wolves were reintroduced, aspen saplings began to recover. This was about more than predators pouncing upon their prey. Plenty of elk escaped the wolves' dinner menu, enough to continue plundering young aspen. But scientists at Oregon State University found that elk were learning where the dangerous spots were, then changing their habit of eating aspen in these vulnerable areas accordingly. "We think these elk need to balance the risk of being killed versus eating in their favorite places," explained forestry professor and study author William J. Ripple. "So, it's a trade-off between food and risk in an ecology of fear."[33]

Climate change adds new challenges to adopting this approach, to thinking holistically about our environment. But direct experiences through travel and citizen science provide valuable insight to overcome those hurdles. Earthwatch volunteers get this. "Doing actual scientific work, collecting data to advance their scientific understanding, really opened my eyes," as my senior colleague Vicky Kleinman explains.[34] Her octogenarian friend Bev Anderson concurs, "There's a spreading information function that goes along with it. I'm sure it changes the volunteers. I've done outdoor things all my life, camping, hiking, river running, but I'm never going to walk through a forest again with the same pair of eyes."[35] We would all be wise to follow her lead.

7

OCEAN TROUBLE

Beware first impressions. As Americans, we often rush our vacation travels, short-changing our cultural and intellectual growth by speeding through visits without fully sampling the unique sights and sounds different locales offer. Seeing may be believing, but make sure your perspective is a complete one, not a partial view that suggests a misleading narrative. Anniversary Reef within Florida's Biscayne National Park, for example, doesn't look like much from afar. Its haphazard splotches of brownish green fail to inspire, at least initially. But taking a longer, closer look, from within the water, reveals an entirely different story.

It's a little chilly at first, not exactly what I'd expected at 80 degrees, water warm enough, as we learned a few chapters ago, for a tropical storm to form. But some goofy underwater calisthenics, even if I look like Woody Harrelson's character in *White Men Can't Jump*, warm me up a bit. Plus, there's the added reassurance of little hurricane danger today. It's mid-April, well before hurricane season traditionally begins on June 1, and the ocean is calm and clear. A cluster of purple common sea fans immediately catch my eye, especially as they contrast with the mustard yellow, burgundy, and burnt orange stony corals below me. But my attention quickly turns to the fish, and I spend the next half hour or so trailing a school led by two brilliant rainbow parrotfish, listening as they feed along the reef. Yes, listening—to a sound I never dreamed of hearing, a parrotfish taking a bite of coral. And yet, these characters regularly graze upon the reef, crunching it up and passing their waste as sand. One parrotfish can create an astounding two thousand pounds of sand in their coral chomping career.

Anniversary Reef is one of more than six thousand individual reefs within the Florida Reef, formed roughly ten thousand years ago when sea levels rose following the last Ice Age.[1] The Florida Reef is the third largest barrier reef in the world, after Australia's Great Barrier Reef and Belize's Mesoamerican Reef. It stretches 360 miles from the Dry Tortugas to the St. Lucie Inlet in Martin County. We are in a portion roughly thirty miles from downtown Miami, in part of the 150-mile-long and four-mile-wide South Florida Reef Tract that attracts more than five million visitors to the Florida Keys each year.

But my son and I are not where most tourists visit. Contrary to popular thought, the Florida Keys do not begin further south at Key Largo. Some fifty additional keys, themselves ancient coral reefs, lie to the north.[2] That is where we are snorkeling today, in Biscayne National Park, the largest marine park in North America. About two-thirds of Florida's coral reef lies within Biscayne National Park and its neighbor, the Florida Keys National Marine Sanctuary, a protected marine area surrounding the better-known portion of the Florida Keys Island chain to our south.[3] Congress designated this watery wonderland of Biscayne a national park in 1980, setting aside 173,000 acres. Before that, beginning in 1968, a much smaller portion was preserved as a National Monument. Ninety-five percent of it is water, including coral reefs and sponge beds, seagrass meadows and mangrove forests, muddy tidal flats and silty bay bottoms. All told, it's home to some six hundred species of fish, three hundred bird species, more than seventy types of marine sponges, forty-five different stony corals, and thirty-five species of octocorals (we'll distinguish corals from octocorals in a bit). Six of the seven sea turtles on the planet also share this space.[4] Throughout, average depth is only six to eight feet, a shallowness that fosters this astounding diversity of life.

But it won't for much longer—unless we act now.

Less than ten percent of the Florida Reef system remains covered with living coral. Back-to-back major bleaching events in 2014 and 2015 hit hard, as Florida Keys Nature Conservancy's Chris Bergh relates. And we now likely face almost yearly bleaching events.[5] "I've definitely seen a difference from when I first started snorkeling," explains Captain Danny Wells as he steers our twenty-three-foot powerboat into a natural channel across Biscayne Bay. "Especially the last five years, there's been a slow but steady decline."[6]

Our six-hour guided trip is with the Biscayne National Park Institute, with two couples besides my older son and me aboard. It takes a little over half an hour to cover the six and a half miles from park headquarters behind the Dante Fascell Visitor Center to our first destination, Jones Lagoon. Next up is an hour and a half paddleboarding through subtropical paradise. The water is crystal clear, and only a couple feet away swims a bonnet head shark, a relative to the distinctive, chill-inspiring hammer heads. But the real stars here are marine sponges and mangroves, both diligently filtering this shallow water so other species can thrive in it. Biscayne National Park, in fact, includes the longest stretch of uninterrupted red mangroves on the eastern coast, with two additional mangrove species, black and white, also found in the park.

Dropping to kneel on our paddleboards, we navigate through a narrow passage leading to Jones Lagoon, with the highly salt tolerant roots of the red mangrove on full display. In addition to their filtering responsibilities, mangroves provide nesting and breeding habitat for fish and shellfish as well as migratory birds and even sea turtles. They are what we call a keystone species, a steady support structure for this entire estuarine community.

Jones Lagoon gets its name from a fellow named Sir Lancelot Jones, who turned down $7 million from developers who wanted to build Biscayne Bay into an industrial powerhouse. Jones, an African American whose father began buying property here with Porgy Key in 1897, opted to sell, instead, with his sister in-law Kathleen, their 277 acres to the National Park Service in 1970 for only $1.2 million.[7] That, combined with a 1968 designation of additional property as a National Monument, ended a decade-long battle between conservationists and developers. Beyond the Jones property, absentee landowners still schemed to transpose this chain of thirty-three keys, formerly a popular pirate hideout and one of the last undeveloped portions of the Florida Keys, into luxury hotels, shopping centers, and beachfront homes. Seven-mile-long Elliot Key was the primary target,[8] but the suggested dredging of a forty-foot-deep canal through the bay to a proposed oil refinery would have threatened the entire region.[9]

Local activists resisted. Articles in the *Miami Herald* stirred interest. Then vacuum cleaner magnate Herbert W. Hoover, Jr., familiar with the region from his childhood, emptied his pockets. Perhaps most notably, Hoover entertained Washington legislators with blimp rides over Biscayne Bay and this same stretch of Florida Keys, thinking the best advocate for the proposed park would be the land and watery landscape itself.[10] Those firsthand travel experiences made a huge difference, generating the needed political support to create first Biscayne National Monument and later Biscayne National Park.

We need this land and water to help do that again. This time, though, the stakes are even higher. While climate change brings many threats, our oceans face perhaps the biggest of all. Less than a third of Earth's surface is land, with oceans covering seventy-one percent. Yet, all too often, as Pulitzer Prize–winning author and environmental historian Jack Davis writes, these seas are seen as only a "passive backdrop."[11] Nothing could be further from the truth, for oceans drive our weather and climate patterns. And the driving life force behind our oceans are corals like those on Anniversary Reef. Corals are akin to trees in a forest, providing both shelter and nourishment. They are a foundation species, providing the nursery that serves a quarter of all marine life. It's not just marine organisms that need coral reefs. Nearly a billion people depend upon the sea for their food, with reefs serving as the foundation for much-needed fishing industries that supply them. Reefs also serve as valued breakwaters for storms, more effective than anything we construct ourselves.

Given all that, let's take a closer look at the coral that constitutes a reef. Hundreds of thousands of polyps, typically no thicker than a nickel, come together to create what we know as a single coral. These corals are what is known as sessile animals because they lack self-locomotion, usually attaching themselves to the reef but able to move with ocean currents. What gives coral their famous vibrant colors, though, are the plant food factories inside them, microscopic algae called zooxanthellae. A coral's circular mouth, surrounded by tentacles, grows over this part plant, part mineral structure. During the day zooxanthellae photosynthesize. This produces oxygen and nutrients for polyps, which return the favor by furnishing carbon dioxide for the zooxanthellae. In time, coral colonies become massive, some even visible from space. But this growth is relatively slow, with individual colonies, depending on the species, only growing half an inch to seven inches a year.

Corals exist almost everywhere in our oceans, from the tropics to the poles. Some seek shallow waters. Others prefer depths of twenty thousand feet. But hermatypic corals, those that build reefs, live only in warm water of 68 to 82 degrees Fahrenheit. This water must also be relatively shallow, less than 150 feet for light to allow photosynthesis. Coral reef development also requires water low in phosphate and nitrogen nutrients, moderate wave action to funnel plankton its direction while dispersing coral waste, and a solid structure to which the initial polyps can attach.[12] The three major types of coral reefs are atolls, fringing reefs, and barrier reefs. The Florida Reef most closely resembles a barrier reef, although it's closer to shore than most and lacks their shallow inshore lagoons. That's why some call it a bank reef.

As noted earlier, there are two main types of coral. There are hard or stony corals like brain and elkhorn coral. And there are octocorals, sometimes called soft coral, such as sea fans and sea whips. Each type holds a distinctive shape, as their names imply, and lives in their own distinct colony. Stony corals are the main architects of the reef, with brain, star, and elkhorn corals, historically, Florida's reef-builders. All are less common now, though, as much threatens them. Culprits include land pollution runoff, fishing, boat-anchors, and even ecotourists like me, if they are careless or uninformed and use reef-toxic sunscreen.

Beyond these proximate dangers, though, climate change itself increasingly assaults the reef, namely with the twin threats of warming water and acidification. At current rates, coral reefs cannot survive our onslaught. Indeed, they will be the first ecosystem we wipe out entirely. A 2012 study, for example, shows half the Great Barrier Reef is already gone. And the prognosis for smaller, more vulnerable reefs is worse. The Zoological Society of London reported over a decade ago that a 360-ppm concentration of carbon dioxide in the atmosphere is the tipping point for long term coral reef viability.[13] We passed 420-ppm for the first time in human history in 2021, according to the Mauna Loa Observatory in Hawaii. Keep in mind the reason these atmospheric numbers matter is that oceans bear the brunt of climate change emissions; they've absorbed 93 percent of all heat tied to extra greenhouse gases in our atmosphere. Our average temperature would be 122° F if oceans did not do so.

Admittedly, coral reefs can withstand higher than normal temperatures temporarily. But only for about a month. If higher temperatures, as little as half a degree Celsius (0.9 of a degree Fahrenheit), extend for two months, most coral species begin to die, first turning a ghostly white. Bleaching is a stress response, and corals behave like people catching a fever. Photosynthesis shuts down, which causes the animal host to expel its food-producing zooxanthellae. This means corals no longer have their primary food source and begin to starve, exposing their malnourished white skeleton underneath. If corals are pure white, they are still alive and not allowing other things to live on them. When they become fuzzy, they are dead.

This is not a problem for Florida alone.[14] The first known global scale bleaching, a planetary heat wave killing coral across the planet, was in 1997–1998, with another in 2010. Such losses add up quickly. Bleaching coral reefs kills off fish populations, damaging human food supply, jobs, and tourism. As Ruth Gates, a marine biologist and past president of the International Society for Reef Studies, warned before she passed away, these consequences will only grow in the years to come. "I always think of the planet as a jigsaw puzzle, and there are all these pieces that must fit together to create the picture," she stated. "Losing pieces like coral reefs or the polar ice cap will ultimately wipe us out as a species."[15]

As bad as this is, coral bleaching is not the only threat our oceans face from climate change. A less publicized threat is ocean acidification. "The ocean takes up quite a bit of CO_2. Estimates are that it takes up about one-third to one-half of all CO_2 emissions to date," explains University of Washington professor Christopher Murray. "It does a fantastic job of buffering the atmosphere, but the consequence is ocean acidification."[16] Two hundred years ago the pH of ocean surface waters was 8.2. Today it is 8.1.[17] That might seem minor, but acidity is measured on a logarithmic scale. That 0.1 difference means that oceans are 30 percent more acidic today than in the early 1800s.[18] That's more acidic than any time in last 800,000 years.[19] And the last time our oceans experienced such rapid change in chemistry was 56 million years ago. That drove many marine species to extinction, with another 100,000 years passing before balance was recovered.

This increase in carbonic acid means calcium carbonate is less available to marine species that build calcium carbonate shells, from tiny sea snails that serve as a foundation for oceanic food webs, to corals, crabs, clams, oysters, and lobsters. And the more acidic water causes those species weakened shells to dissolve, leaving them vulnerable to predators and disease. This pH shift also changes how sound waves travel through water. With a 0.3 decrease in pH, sound travels 70 percent further in water, so our oceans are becoming louder, likely confusing animals like cetaceans that communicate through sound. Acidification means smaller fish, too. Previous researchers assumed fish were mobile and tolerant of heightened CO_2 levels. But University of Washington professor Christopher Murray and University of Connecticut marine scientist Hannes Baumann, studying shorter living fish across their life cycle, found that high CO_2 concentrations in water can make fish grow smaller.[20]

All told, you don't have to be a coral reef specialist to understand that our oceans are in trouble. But we can find solace in one. "It's not too late for coral reefs," states Ove Hoegh-Guldberg, professor of marine studies at the University of Queensland in Brisbane, Australia, and a climate specialist on coral reefs. "It's still possible to reduce the rate at which the climate is changing, and that's within our power today." The chips may be down, as Hoegh-Guldberg says, but if we act within this decade, if we build resilience within the system, we can save a substantial portion of reefs across the globe.[21] Biscayne National Park is a powerful medium to achieve that end. Seeing this special landscape convinced Americans to save it over half a century ago. Witnessing the climate threats unfolding within it now could well reprise that role today.

8

HEAT AND HEALTH

Anchorage hit a new high yesterday, reaching 90 degrees for the first time. That might not sound terribly hot if you are from Phoenix or Miami, but Anchorage is a different story—and this record breaks a previous peak, set over half a century ago, by five degrees.[1] Heat's not what comes to mind when picturing icy Alaska, but, alas, increasingly, it should be. It's the day after the Fourth of July, and the smell of smoke in our hotel hallway is strong. This haze isn't from patriotic fireworks celebrations last night, though. "Extreme dry weather conditions" forced the Anchorage Fire Department to cancel all official shows. And across much of the state, fireworks sale and use were banned earlier in the week. We are smelling, and breathing, smoke from wildfires, a traditional fixture in Alaskan summers. But because of climate change, these summer months are starting sooner—and lasting longer. That means wildfire season is roughly 40 percent longer than in the 1950s, according to the nonprofit group Climate Central, with thirty-five additional days stretching now from May to early August.[2]

Rising temperatures are to blame, with higher heat soaking up more moisture and drying out the landscape. Heat records like this will likelier be more severe for roughly 80 percent of our planet in the coming years, according to a team of researchers led by Stanford climate scientist Noah Diffenbaugh.[3] Alaska stands out here, for the state is warming two and a half times faster than our lower forty-eight. That speeds up snow and ice melt, with snow no longer found on the ground in some parts of Alaska two weeks earlier than historically expected.[4] This is important because, as noted in chapter 5, melting snow and ice leave behind less reflective, darker snow and water. That accelerates the warming process even more.

It's not just Alaska we must worry about. It is "very likely that heat waves will occur more often and last longer, and that extreme precipitation events will become more intense and frequent in many regions," according to the 2014 Intergovernmental Panel on Climate Change Summary for Policymakers. By 2100, temperatures will likely range from 4.5 to 14 degrees Fahrenheit (2.5 to 7.8 degrees Celsius) higher when including climate uncertainty and no additional mitigation.[5] Such impacts will be even more harsh after considering something called the urban heat island effect. This is the impact asphalt and lack of shade have on temperatures within cities. Dark asphalt absorbs heat during the day. At night, it releases that heat, raising local temperature as much as 22 degrees Fahrenheit. With four out of five Americans living in urban settings, this is a growing problem, restricting economic productivity, draining energy resources, and endangering human health.

It's only going to get worse.

Food and water will be harder to supply. Disease will be more common and tougher to contain. Illnesses will be more frequent and less well treated. And more people will die. Heat stress already claims over 150,000 lives worldwide each year. Those numbers will likely more than double between 2030 and 2050, according to the World Health Organization. "Heat is the mothership of climate risk—it exacerbates hurricanes driven by warmer air and warmer water, drought and desertification, devastating and cataclysmic wildfires, food insecurity, water shortages, increased violence, and immense economic loss," states Kathy Baughman McLeod, the senior vice president and director of the Adrienne Arsht-Rockefeller Foundation at the Atlantic Council.[6] "And it's not just extreme heat," McLeod adds. "It's the slow-roasting increased nighttime temperatures that don't allow the body to cool down and repair itself. We are literally cooking ourselves."

Heat is already killing more people across our country than floods, hurricanes, lightning, tornadoes, and earthquakes combined. These deaths by heat are likely underestimated, too. As a "silent killer," the casualties heat inflicts are often officially recorded as something else like kidney failure or a heart attack. Researchers at the University of British Columbia School of Population and Public Health, along with the Boston University School of Public Health, believe that heat may cause thousands of American deaths each year, far above the six hundred estimated by the Centers for Disease Control and Prevention.[7] "How dangerous a hot day is may depend on where you live," elaborates study lead author Kate Weinberger, assistant professor of occupational and environmental health at British Columbia's School of Population and Public Health. "A 32-degree Celsius day might be dangerous in Seattle, but not in Phoenix. One of the factors that gives rise to this phenomenon is differing degrees of adaptation to heat. For example, air conditioning is much more common in cities like Phoenix that experience hot weather frequently versus cities like Seattle with their traditionally cooler climates."[8]

Keep in mind these impacts are uneven within a city, too, hitting those less fortunate the hardest. Most notably, heat disproportionately affects black and brown communities as racist urban-planning policies expose these populations more intensely to the urban island heat effect. A key driving force here was something called the Home Owners Loan Corporation (HOLC), created during the first one hundred days of President Roosevelt's administration in 1933. It was welcomed then, by white America, as part of the New Deal and our nation's struggle to prop up housing during the Great Depression. Using data and evaluations from local real estate appraisers, lenders, and developers, this body graded residential neighborhoods with visually striking color-coded maps. Those deemed safe investments received an A-grade and were colored green, encouraging mortgage lenders to extend loans on these properties. Those receiving the lowest D-grade were labeled "hazardous," colored red, and essentially abandoned by whites.[9]

There are 239 HOLC maps altogether. Not all of them were finished, and they exist only for cities with forty thousand or more people. But the story they tell is visceral.[10] Historian Kenneth Jackson rediscovered them in the National Archives in the 1980s. As he writes in his award-winning book *Crabgrass Frontier*, HOLC "devised a rating system that undervalued neighborhoods that were dense, mixed, or aging" and "applied [existing] notions of ethnic and racial worth to real-estate appraising on an unprecedented scale."[11]

In other words, they were racist.

Federal housing policy formally blocked African Americans from acquiring real estate capital. Seeing these neighborhoods today, almost a century later, the differences are stark. And they highlight what sociologists call social vulnerability. Robert Nelson, an expert on housing policy and race at the University of Richmond, and coauthor of a digital mapping project that spells out this history,[12] explains this to me one late April morning at his home. "There' a calcification of privilege and poverty," says Nelson. "People's jaws often drop when they see the maps. There's recognition that looks like my city today. Segregation persists. You can see the same neighborhoods. Not a lot has changed from the grades then to today. And the other thing you show them are the area descriptions, which are so clearly racist. It's incredibly racist throughout, and nativist, and anti-Semitic, and that brings it home. How systemic racism created investment in some areas and divestment in others."

Redlining nearly ninety years ago shaped the demographics of urban America today. And that racism, magnified by climate change, continues to threaten the health of black and brown communities across our country.[13] "Any market is partly based on psychology," continues Nelson. "The argument is white people are worried African Americans are going to settle in their neighborhood and their home values will go over a cliff, just plummet. What HOLC does is say the federal government is not going to let that happen. So, you don't worry about buying a house in that neighborhood. It's doubling down on systemic racism as a way of bolstering confidence in the real estate market for well to do white people."

Racial segregation in housing became illegal with the Fair Housing Act of 1968. But its legacy persists today in cities like Richmond and across this land, with dangerous, and at times deadly, consequences. Research by climate scientists like Jeremy Hoffman, chief scientist at the Science Museum of Virginia, sheds additional light here. As he demonstrates, those living in communities without tree canopy and native greenspaces are at higher risk for heat-related illness and ailments. Ambulance call rates are higher. Urgent care center check-ins are higher. And asthma rates are higher.[14]

Other racist influences, like where polluting industries set up shop, cannot be discounted. But drastic differences in urban heat, from neighborhood to neighborhood, cannot either. Hoffman and his coauthors, Vivek Shandas, professor of urban planning at Portland State University, and Nicholas Pendleton, now a research assistant at George Washington University's School of Public Health, used the digital mapping work of Nelson and his colleagues to demonstrate this. Their work highlights the effects of historical housing policies on those living in over one hundred urban areas.[15] "We can see a pattern in a city, and it all seems to relate back to this urban heat island effect," states Hoffman. "There has been structural disenfranchisement in certain neighborhoods."[16] "Through an analysis of historic redlining maps and modern satellite imagery, we've found a relationship between government-led housing segregation of the past century and vulnerability to extreme heat and flooding in Richmond today," he adds.[17]

We've known it's getting hotter for quite some time. A 1988 heat wave first caught national attention.[18] People felt its prolonged effects throughout the eastern two-thirds of our country. Further west, Yellowstone National Park infamously caught fire, burning over 793,000 acres from mid-June to mid-November. Congress even held public hearings. Colorado Senator Tim Wirth invited an array of climate experts to testify before the Energy and Natural Resources Committee, first and most famously climate scientist Jim Hansen, then director of the NASA Goddard Institute for Space Studies. "The Earth is warmer in 1988 than at any time in the history of instrument measurements," Hansen declared at that congressional hearing on a blistering 98-degreee June day in Washington, DC. "It's time to stop waffling so much and say that the evidence is pretty strong that the greenhouse effect is here."

Nearly thirty-five years later, we now understand this warming is decidedly uneven. Temperature rises will be more significant in higher latitudes than equatorial areas. They will be greater over land than oceans. And higher within the interior of continents than along their coasts. That's what makes yesterday's 90-degree record in coastal Anchorage, lodged between the Chugach Mountains and Cook Inlet in the Gulf of Alaska, so notable. Summer temperatures in the Alaskan interior can reach 95 degrees in June and July, but not along the coast, at least not until now.[19] While it's not quite 90 in Anchorage the next day when we arrive, it's still twenty-five degrees above normal July highs of 62.

Across this land, especially here in Alaska, summer now means something profoundly different, as environmental author and activist Bill McKibben predicted more than three decades ago.[20] These rising temperatures present a range of threats. And these threats often interact with one another, compounding damage, even when they first develop on opposite sides of the globe. In part, that's why the traditionally conservative U.S. military recognizes climate change as a dangerous "threat multiplier."

Consider basic economic productivity. The optimal annual average temperature for economic productivity, according to three researchers from Stanford University and the University of California, Berkeley, is 55.4 degrees Fahrenheit.[21] Conveniently, that's our historical median here in the United States, as well as several other large world economies.[22] As heat waves linger, though, productivity will decline. Utilities will struggle to supply enough power to the electrical grid. Brownouts will be more common. And air conditioners will run more frequently, creating an unwelcome positive feedback loop.

Air conditioners and fans account for 10 percent of global electricity consumption to date. If powered by fossil fuels, more usage translates into more greenhouse gases released into our atmosphere. With air conditioners, moreover, there is the additional release of hydrofluorocarbon (HFCs) chemicals that also cause warming, trapping thousands of times as much heat in the atmosphere as carbon dioxide, molecule for molecule. This release occurs at several stages, during the manufacturing process, when equipment is improperly disposed, and with leaks while air conditioners are in use. On top of all that, there is the heat waste ACs deposit even when not leaking. "There's a lot more waste heat being dumped into the environment from their attempt to keep buildings cool," explains David Sailor, a professor at Arizona State University and the director of its Urban Climate Research Center. "That creates a kind of positive feedback loop between local heat and global climate change."[23]

These are not trivial numbers. While air conditioners use only about 6 percent of all electricity in the United States, that number doubles to 12 percent for home energy expenditures, according to the U.S. Energy Information Administration. In hot and humid parts of the country, like Florida and Texas, the annual average jumps to 27 percent.[24] And looking expressly at summer months, our nation's warmest climates devote well over 50 percent of their home energy usage to air conditioning. As temperatures rise, so will these energy needs. Thinking beyond the United States, the numbers are even worse. Air conditioning in India, for example, consumes 40 to 60 percent of the country's electricity, despite less than 10 percent of homes having any units. As more homes acquire air conditioning, that number will inevitably rise. "This is going to matter a lot," says Lucas Davis, an associate professor at the Haas School of Business at the University of California Berkeley. "If one is thinking about energy and environment in the next couple decades, you have to think about cooling."[25]

Our infrastructure is not built for these rising temperatures, from warping train tracks known as "sun kinks" in Illinois or New York to buckling roads after permafrost melt in Alaska. Even air traffic is vulnerable. Sometimes it's simply too hot for smaller jets to take off.[26] Hotter, thinner air limits their ability to generate enough lift, a scenario that canceled more than fifty flights in Phoenix in June 2017. "When you get in excess of 118 or higher, you're not able to take off or land," explains Ross Feinstein, a spokesman for American Airlines.[27] Radley Horton, a research scientist at Columbia University's Earth Institute, published a joint study with PhD student Ethan Coffel on the effect of extreme heat on aviation two years before the Phoenix airport groundings. They predict possibly four times as many weight restriction days for our most at-risk airports by 2050. "To take off, a plane has to reach a certain minimum speed," states Horton. "On hot days and at high elevations, that minimum speed increases. High elevation and high temperature mean less molecules of air for the plane to push off of."[28]

Not all airports are affected equally. High-altitude Denver, with its thinner air, is more vulnerable to high temperatures. New York City's La Guardia airport, while at sea level, has a relatively short runway compared to other major commercial airports, so it, too, will face cancellations. Reagan National Airport in Washington, DC, will as well thanks to its similarly short runway. On top of that, even a no-fly window lasting only an hour or two at just one of these airports can create considerable ripple effects across the rest of the country.[29]

Rising temperatures, as we saw a few chapters ago, are also melting our cryosphere. This depletion of glaciers initially increases river flows, providing more water for irrigation and agricultural production—but only temporarily. Such snow is the leading source of irrigation and drinking water for many agricultural regions across the globe, including the southwest United States. Take the Colorado River, for example. It's that region's primary source for irrigation, depending in large part on snowfields in the Rockies. California also depends heavily on the Colorado, along with area snowmelt from the Sierra Nevada Mountain range, to irrigate its Central Valley, our nation's fruit and vegetable basket. Alas, the Sierra snowpack, providing about one-third California's fresh water annually, is predicted to continue to shrink by perhaps 54 percent in the next two to four decades.[30]

In addition to this melt, there is less new snow each year. As temperatures warm during the winter, and spring storms come earlier, more precipitation falls as rain instead of what was previously snow. "We are currently relying on snow to store the precipitation we get in the winter for use during the summer," said Andrew Jones, a climate scientist at the Lawrence Berkeley National Laboratory, who worked on the study. "In the future, we'll have to radically restructure the way we manage water in California."[31] As an aside, while we've already seen how earlier snowpack melt gives higher temperatures more time to dry out a forest and lengthen fire seasons, drought and fire are not the only climate risks agriculture faces.[32] Increased carbon dioxide allows plants to grow faster and larger, but, without additional nutrients and minerals, these crops lose their relative nutritional value. Food will be less healthy, then, in our warming world.

Hotter temperatures also mean more bugs are coming. Rising temperatures affect insects in two key respects. For one, it boosts their metabolic rate, making them hungrier for plants. And secondly, their life cycles speed up, causing bugs to reproduce more quickly. "The warmer readings could mean the insects will be capable of producing up to three generations of their kind in a single growing season, filling fields with their hungry offspring," warns Stanford climate scientist Noah Diffenbaugh.[33] Not only will numbers increase in areas they've previously frequented, but insects will expand their traditional range as well. The European corn borer, Western corn rootworm, and corn earworm are all expected to carve out new territory in our warming world. This may well push farmers to use more pesticides, causing further complications.

Bugs also impact human health with the diseases they carry. Blacklegged ticks, for one, transmit Lyme disease, whose symptoms range from joint pain and fatigue, to memory loss and facial palsy. Even years after treatment, symptoms can continue. These ticks require precise temperatures, rainfall, and humidity that historically restricted their range to eastern and southeastern United States. But climate change is likely extending their range, along with the Lyme disease they carry.[34]

Climate change also creates more habitat for one of the most prolific set of killers in history, mosquitoes. As temperatures warm and our climate becomes more humid with heavier rainfall, mosquito season is getting longer. Interestingly, higher temperatures also reduce the size of larva and subsequent adults. These smaller adults must feed more frequently to develop their eggs, which increases chances of transmission of the diseases they carry. Warmer temperatures have already extended the geographic range of the West Nile virus. And with the Dengue fever virus, the time spent incubating inside a mosquito is shortened when the mercury rises. The Zika virus also appears to benefit from our warmer world. Historically limited to Uganda and southeast Asia, it did not appear to cause birth defects until recently. Microcephaly, the condition where a baby's head is much smaller than expected, seems to have arisen only as Zika spread to the Americas. And then there is malaria. The world's fourth leading cause of death in children younger than five years, malaria kills over one million people each year. Climate and land-use changes are expanding its traditional range beyond sub-Saharan Africa, with an estimated 40 percent of the world's population now living in malaria-risk areas, according to the World Health Organization.[35]

Bugs are not the only threat to human health as temperatures rise. A warmer world creates opportunities for pathogens to mutate, to jump to new hosts, and to travel outside their traditional geographic locales. Hotter lakes and oceans will also foster more bacteria, parasites, and algal blooms.[36] Higher temperatures bring heavier rainfall, sparking sewage overflows and contaminating drinking water. That's an ideal environment for cholera and other waterborne diseases.[37]

Heat stress also challenges and creates risks for our vital body organs. Blood flow must increase to carry heat out of the body when temperatures rise. That means the heart must pump harder and faster, which can cause heat exhaustion, heat stroke, and even heart attacks.[38] "Cases of cardiovascular disease are likely to rise, due to hotter temperatures and an increase in extreme weather, says the National Institute of Environmental Health Sciences. "Extreme hot or cold temperatures cause physical distress, which may increase the demands on the heart muscle and prevent the blood supply from doing its job, triggering a heart attack," explains cardiologist Farhad Rafii.[39] Still another negative impact is that heat exacerbates chronic obstructive pulmonary disease (COPD), a chronic inflammatory lung disease, by affecting pressure in the lungs and aggravating asthma. Heat damages other organs, too, such as the liver and kidneys. Twelve senior citizens died from various organ failures in a south Florida nursing center, for example, when Hurricane Irma knocked out the facility's air conditioning in September 2017—and temperatures inside soared to a sweltering 99 degrees for more than two days.

Additional hot days also translate into a longer allergy season as flowers and trees bloom earlier and longer. Ragweed pollen season, for example, has increased between eleven and twenty-seven days in the United States and Canada. It's not just increased duration of pollen season that's a problem. More pollen is produced as well. Ragweed, for example, grows 10 percent taller and produces 60 percent more pollen as temperature warms, according to one Harvard study.[40] As anyone with allergy induced asthma can tell you, this is more than a mere inconvenience. "You feel like you can't exhale, and you can't breathe in enough," says Cheryl Holder, an internal medicine specialist at Florida International University and president of the Florida State Medical Association.[41]

We've become all too familiar with the globalization of disease these last couple years during the coronavirus pandemic. As of this writing, COVID-19 has infected more than ninety-two million people in the United States, claiming more than one million American lives alone. But with scientists suggesting possibly more than one million viruses still undiscovered, future pandemic challenges likely lie ahead. Key demographic factors shape how vulnerable we will be. With rising temperatures, risks are highest for those already susceptible to heat, from older adults and children to pregnant women and outdoor workers. Rural communities, where livelihoods are often interconnected with agriculture and long hours outside required, are at risk. But, as noted earlier, so are urban areas, where the urban island heat effect can be downright deadly. Communities of color and low income face this already. They will even more in years to come.

"For too long, a lot of the climate change and global warming arguments have been looking at melting ice and polar bears and not at the human suffering side of it," states the father of environmental justice Robert Bullard, professor at Texas Southern University. "They are still pushing out the polar bear as the icon for climate change. The icon should be a kid who is suffering from the negative impacts of climate change and increased air pollution, or a family where rising water is endangering their lives."[42] In short, a warming world is dangerous for all, but those less fortunate will bear disturbingly disproportionate harm. Opening our eyes to the dangerous inequities within our own communities is the first step to correcting this environmental injustice. But taking concrete actions must follow, so let's turn next to what we can do to make a difference.

II

DO IT YOURSELF

Action Making a Difference

9

HERE COMES THE SUN

"Here Comes the Sun" resonates reassuringly across the green and gold campus auditorium, the old Beatles tune increasingly competing with a rising din of excited conversation. The atmosphere resembles a political rally, and solar energy is taking center stage. Those assembled at the third annual Florida Solar Congress are largely white and older like me, a demographic deficiency to which I will return. But we are all here on the University of South Florida campus for one reason: promoting solar energy, particularly distributed solar on rooftops like yours and mine.

Solar power is our most accessible renewable resource. It's our largest, too. The sun provides, in an hour, roughly the same amount of energy our entire planet uses in a year. Solar power is intermittent but predictable. And it is more evenly distributed than any other form of renewable energy. Two types of solar technology exist, solar thermal and photovoltaics. Solar thermal panels convert sunlight into heat and offer the distinct advantage of storing energy. They are the more efficient option, around 70 percent, but limited to heating fluids, namely water. Think swimming pools and hot water heaters. On the other hand, photovoltaics is the production of electricity by exposing material to light. It registers a much lower efficiency, ranging typically between 15 and 22 percent. But PVs ability to generate electricity, not just heat water, has cemented its leading residential role today. It's where the action is.

PV history dates to its discovery in 1839, although the first solar array was not installed until decades later in New York City in 1884. Over the years, PV technology has had its share of renowned scientific patrons. Albert Einstein won the Nobel Prize for Physics with his explanation of the PV effect in 1921. And in 1931 Thomas Edison famously exclaimed to Henry Ford and Harvey Firestone, "I'd put my money on the sun and solar energy. What a source of power! I hope we don't have to wait until oil and coal run out before we tackle that."[1]

The modern solar era began, by accident, in 1952 when three scientists at Bell Labs in Princeton, NJ, discovered that sunlight striking silicon-based material produced electricity. At first, materials were so expensive their only practical application was with satellites or remote scientific research from far reaches of our planet. But costs, continually outpacing predictions, fell remarkably over the last half-century, from over $70 per watt in the 1970s to $3.50 per watt in 2000 to $2.49 in early 2021, although this number rises or falls depending on which state you live in.[2]

Solar PV has many advantages. It's most available when most needed, on hot summer afternoons. It's sustainable. It's easy to install. It's surprisingly scalable. It operates noise-free. And without moving parts, it's essentially maintenance-free. That keeps costs down—and helps systems last longer. Although Florida sun is notoriously tough on roofs, even our panels perform at 80 percent efficiency for twenty-five years. No damage is done during installation, by the way. Solar panels actually protect roof shingles, prolonging lifespan and lowering AC demand. And of course, once installed, PV is greenhouse gas emission free. Even if that's not on your radar, or your political persuasion leans right, there's good reason to support solar. It allows production of your own power. It offers independence.

While several Florida cities ironically receive more rain on average than notoriously dreary Seattle, the Sunshine State still ranks high in sun potential. Southwestern states have the best solar resources in our country, among the best in the world. But Florida boasts more sunshine than any state east of the Mississippi River, 230 days to be exact. Only California and Texas hold more rooftop solar power potential than Florida, according to the Department of Energy. And Florida is fourth in the union when it comes to installed capacity, according to the Solar Energy Industries Association.[3]

Florida's rise in the rankings the past decade is thanks to grassroots activists like those this Saturday morning—and their nonprofit host, Solar United Neighbors (SUN). The SUN story is inspiring. It began in 2007 when two young boys, twelve-year-olds Walter Lynn and Diego Arrene-Morley, saw Al Gore's documentary *An Inconvenient Truth.* Motivated to make change themselves, they agitated for rooftop solar on their own homes. Walter's mother, Anya Schoolman, was sympathetic but quickly discovered a complicated maze of logistics, not to mention intimidating expenses. She suggested that the boys gauge neighborhood interest, thinking, together, their block might bargain for a better price. After two weeks of flyer/questionnaire distribution, Walter and Diego had fifty households interested, with forty-five eventually signing solar contracts. News of their Mt. Pleasant Solar Co-op spread throughout the district, and other neighborhoods requested advice. Two years and hundreds of meetings later, DC Solar United Neighborhoods (DC SUN) was born. Today it totals thirteen states plus that original DC chapter. My adopted state of Florida is one of them.

Mary Dipboye, co-founder of Solar United Neighbors of Florida, traces our Sunshine State solar co-op origins to an NPR story she heard about that DC nonprofit in 2015. Like Walter and Diego before her, she was inspired. "One woman accidentally did something," she thought. "I wonder if I could do that?"[4] Dipboye went to her church and asked if anyone was willing to help. A guy named Michael Cohen raised his hand and, together, they started two co-ops in Orlando that same year. That success led to a statewide program the following year.

Here's where Mary and Michael's experience, like Walter and Diego's before them, offers a powerful lesson. All four were committed activists, willing to devote precious time to the solar cause. But they also realized their work would be much more impactful if they brought others onboard. In Florida, that meant reaching out to established organizations, including a fruitful partnership with the League of Women Voters of Florida. With thirty-one chapters, the League offered experienced media outreach and an opportunity to better tell the solar story. Dipboye credits that step as the game-changer, the stage at which her individual actions with Cohen scaled up to a make an even bigger difference.

We need more of that type of thinking when it comes to climate action.

It's a Wednesday evening, and a newly formed East Orange County Co-op is hosting the first of two information sessions to recruit new members. I've always been curious about rooftop solar but found the technical components intimidating, especially given its initial price tag. Having spoken with Dipboye months ago, though, I'm less intimidated now. It's a tough crowd. About one hundred people, not counting the baby babblers tagging along with their progressive parents, pepper presenters with a mix of technical and financial questions. There's a lot to absorb, but I'm thinking this might be possible. Dinner afterward with Dipboye's colleague Michael Cohen seals the deal. My motivation was initially environmental but has become economic as well. "That's normal," says solar activist Alan Brand. He finds a 50-50 split in motivating factors for those that commit to rooftop solar. "Some people need just a little nudge to the other side, not the dark side but the sunny side," he chuckles through the telephone.[5] Brand should know. After attending a solar fair much like the one here at USF, he was recruited to speak at the Dunedin Public Library. Brand designed his first ninety-minute talk out of frustration with misconceived notions about local solar power. Since then, his talk has continually evolved, sometimes lasting two hours, but always focusing upon financial aspects while borrowing heavily from his Rochester Institute mechanical engineering degree.

Solar photovoltaic is not complicated. Typically made from silicon, the main component in natural beach sand, solar cells are combined into three-by-five-feet panels the size of an American flag. Those modules are then mounted together onto a racking system of metal bars that connects easily to standard asphalt shingle or metal roofs. Power is produced as direct current (DC), and then converted for home use to alternating current (AC) by an inverter. Electricity production is measured in kilowatts (KW), comparable to the miles per gallon we use in characterizing automobile efficiency. Each year efficiency improves. Average system sizes in Florida are around 8 KW, which equates to roughly a $20,000 base price.[6]

That is where obstacles arise: the price point. Twenty thousand dollars is not chump change. Plus, in states like Florida, all those costs come upfront. You are essentially prepaying almost a decade of energy bills. Going solar is a major purchase, like buying a car. And as Dave Pearce, former founding CEO of thin-film solar cell maker Miasole, adds to this analogy, that includes paying upfront for nearly a decade of gasoline, too.[7] Some homeowners can afford that. Many can't.[8] Housing sectors beyond the traditional detached homeowner are also a challenge. Michael Cohen, Dipboye's cofounding colleague of SUN FL, hopes to change that, incentivizing condominiums and rental properties to adopt solar co-ops as well. That would be huge, not only expanding solar production but enlarging the constituency supporting it.

That's key. As Brand noted, environmental concerns motivate around half solar converts. But just as many are driven by financial interests. After its eight-to-ten-year payoff, solar PV provides free electricity. "It's like printing money on your roof," Cohen grins. "Every time you go to open your utility bill is like Christmas."[9] And by the way, utility electricity prices for the nonsolar will not stay constant in the years to come. Energy experts predict 2 percent annual increases on average, so solar saves even more money over time.

Homeowners, of course, are not the only solar game in town. Two other types of solar application exist, utility-scale solar farms and something called community solar. Environmental entrepreneur and activist author Paul Hawken's extensive compilation of greenhouse gas mitigation measures, *Drawdown*, ranks utility-scale solar farms as eighth on its list (distributed solar is tenth). These farms first sprouted in the early 1980s and are found today in deserts, on military bases, atop closed landfills, and even floating on reservoirs. Compared to rooftop solar, solar farms enjoy lower installation costs per watt. Their efficacy in translating sunlight into electricity is also demonstrably higher. Rotating panels, for example, make the most of sunrays and improve efficiency by 40 percent. All that helps explain why climate change mitigation, for years to come, will owe much to utility-scale solar farms and their continued growth.

But considerable room and political promise remain for our other two arenas of application, distributed rooftop solar and community solar. Community solar, for example, allows consumers to enjoy the mix of economic and environmental benefits from going solar even when utility providers are slow to switch. Because the array is shared with others, you don't own and operate these panels yourself. You purchase solar shares or a subscription from a third party. Community solar is large-scale, privately owned infrastructure that creates opportunities for those unable to go solar on their own, whether due to lack of space, limitations from a northerly facing roof, too much shade, HOA or condo restrictions, or inability to afford those high upfront expenses. For a handful of states, unfortunately, the sticking point here is that this energy is not produced by utility companies but what are called third party providers. Some states, such as Minnesota, are leading the charge when it comes to community solar. Others, like Florida, are not.

The reason is simple. Third party providers are a threat to the utility industry's century old business model, one rooted in centralized control over the entire electricity system, from generation to transmission to distribution. More on that in a minute. For the Sunshine State to truly reach its potential, it must become, paradoxically, more like chilly Minnesota. Florida is one of four states (Kentucky, Oklahoma, and North Carolina are the others) that bans third party sales, what are known as Power Purchase Agreements (PPAs) in the industry. Solar advocates tried to change that law it 2016 with an amendment to the Florida constitution. They were unsuccessful—but it could have been worse. Disaster was narrowly avoided when a counter amendment designed to crush the rising popularity of rooftop solar was voted down. "As a coalition representing every part of Florida's political spectrum, we defeated one of the most egregious and underhanded attempts at voter manipulation in this state's history," Tory Perfetti, chairman of Floridians for Solar Choice and director of Conservatives for Energy Freedom, joyfully declared that election night. "We won against all odds and secured a victory for energy freedom."[10]

Perfetti trumpeted the amendment's failure, and rightly so. But it's important to remember the original goal for solar activists was much greater. Floridians for Solar Choice was an eclectic coalition led by the Southern Alliance for Clean Energy (SACE). It included green Tea Party activists, environmentalists like the Sierra Club, Florida's retail and restaurant federations, and religious groups such as the Christian Coalition of America. Together, they sought to place their own amendment on the ballot first, one that amended the state constitution to end utilities' monopoly on retail electricity sales. It would create more choices, allowing consumers to install leased solar panels on rooftops at no upfront expense. That pro-solar amendment's outlook was initially bright, with polling support around 70 percent. Utilities adjusted adeptly, though, and created a faux grassroots group called Consumers for Smart Solar. They started their own ballot measure and spent more than $25 million, with key contributors such as Florida Power and Light, the state's largest electric utility, and Duke Energy, Florida's second-largest utility.

To be successful, amendment campaigns typically hire paid petition gatherers, paying per signature. The utility-backed measure paid their gatherers twice as much per signature as the true solar advocates. "When we were paying a dollar on the street, they were paying $2," explained SACE head Stephen Smith. "When we were paying $2, they went to $4." With some 700,000 signatures needed for acceptance on the Florida ballot, solar proponents were spending $350,000 a week to meet their target.[11] They simply could not afford to continue.[12] Faced with this predicament, Floridians for Solar Choice realized they needed to change tactics. They abandoned their own amendment and pivoted focus to a No Campaign against the utility proposal that did make the 2016 ballot, deviously misleading amendment 1. Using seemingly pro-solar rhetoric, it merely affirmed a preexisting consumer right to own solar.[13] Proponents targeted Florida's notable elderly population, claiming it would protect seniors from scams. They also targeted black and Latino communities, suggesting that third-party solar would raise electricity rates on poorer populations. And its wording masked how the amendment would stymie the rooftop solar market by creating "constitutional protection for any state or local law ensuring that residents who do not produce solar energy can abstain from subsidizing its production."[14]

The target here was net metering. Net metering allows electricity to flow both to and from the customer. When generation is more than use, customers send that excess electricity to the utility. Those credits are banked for subsequent months or purchased from consumers at the end of a year, albeit often at wholesale instead of retail rates. Some forty-four states have net metering policies, although nineteen of them, pressed by concerned utilities, are considering changes. "The wolf is always at the door when it comes to net metering," warns George Cavros, a Fort Lauderdale lawyer who works for the Southern Alliance for Clean Energy.[15]

Utilities argue rooftop solar homeowners are free riders who exploit the utility grid without paying when they use net metering. One can just as easily argue the opposite, though, that rooftop producers pay for their own equipment and volunteer their real estate, which reduces what utilities need to spend to produce electricity. In effect, solar homeowners subsidize those without solar, particularly as the rate we receive for producing additional electricity is far less than what utilities charge. Those differences aside, the anti-solar measure was set to pass, polling at 70 percent according to Scott Thomasson, director of new markets at Vote Solar, if not for a fortuitous slip of the tongue. When the *Miami Herald/Tampa Bay Times* reported on a leaked audio recording of Sal Nuzzo, a policy director at James Madison Institute, the tide began to shift. Hired by a fake grassroots solar group called Consumers for Smart Solar, Nuzzo was speaking at a conference held by his think tank when he referred to the Florida initiative as "a little bit of political jiu-jitsu" worth emulating.[16] His comments turned out to be the ammunition solar advocates needed, dropping election day ballot support to 50.8 percent (60 percent is the threshold Florida amendments need to pass).

That brings us back to square one in Florida when it comes to third party providers. And even worse, utility companies still see distributed solar as a threat, one that turns customers into competitors. In late 2021, for example, they tried a different route, with Florida Power & Light drafting a bill for their lobbyist to deliver to state Senator Jennifer Bradley and Representative Lawrence McClure, one that would end net metering in Florida. By early March 2022, the state legislature passed that bill, an action that would have killed investment in distributed solar here if not for Governor Ron DeSantis. Quite unexpectedly, the governor vetoed the bill passed by his fellow Republicans and upheld our pro–net metering policy, his political position shaped, in part, by an overwhelming grassroots response. Individual citizens made sure their voice was heard, with 16,809 emails, letters, and phone calls flooding DeSantis's office to urge overturning the unfriendly solar bill, compared to a mere thirteen in support of it. Old-fashioned grassroots activism made a difference.

The truth is utilities' centralized model worked well for much of the twentieth century, spreading electricity access into rural and remote areas that otherwise would not have received it. But times have changed over the last hundred years. Strong arguments exist for decentralizing power in the twenty-first century, ranging from political arguments centered around consumer choice, to economic rationale encouraging market competition, to national security concerns about reducing centralized vulnerability from terrorist threats. And of course, there is the climate change piece to our puzzle.

Solar United Neighbors co-op system, with its emphasis on local production, addresses all the above. Co-ops are free to join, with no obligation to purchase panels. Once there are thirty participants, a co-op selection committee picks the installer, basing their decision upon a mix of price, equipment quality, and warranties, along with added consideration for local companies. Homeowners sign contracts individually, but larger numbers within the co-op allow installers to order in bulk and save on materials costs. This encourages them to bid at a lower price than they would otherwise. That often equates to a 20 percent installation discount says solar activist Rick Garrity, who retired in 2015 after fifteen years as executive director of the Environmental Protection Commission in Florida's Hillsborough County.[17]

Keep in mind, installers are not losing money by working with co-ops. Their profit margins remain strong thanks to savings in other areas. Finding customers costs money. Since soft costs on advertising represent roughly 15 percent of total solar price, landing a co-op bid reduces expenses. The industry average on closing a solar deal, from lead to sale, is 10 percent. SUN routinely doubles that, producing above 20 percent returns, relays Florida SUN program associate, Heaven Campbell. That higher conversion rate translates into higher installer profits.

Back on the consumer end, co-ops also offer a wealth of solar technical expertise, including assistance in engaging with installers. "I hear from a lot of people that they don't know who they can trust, don't know who is reputable and who is not," says realtor Lynn Nilssen, a co-coordinator of the Sarasota Ready For 100 Committee, which seeks 100 percent renewable energy citywide by 2045 and for city buildings and operations by 2030.[18] Uncertainty is an often-understated handicap to going solar. Money matters, but so does consumer comfort level with the product being purchased—and from whom it is purchased. Potential solar co-op participant Oscar Vargas agrees. "The biggest obstacle [to me] is the complexity of information available," says Vargas. "There's so much information; it can be overwhelming . . . It takes time and commitment to go through the process, to do your homework."[19]

Co-ops do that work for you. Co-ops also create opportunities to connect with fellow solar enthusiasts, to become part of a growing solar movement. Building community addresses the all-important scalability question, but directly through people instead of solar panels. "The coop model resonates [by] generating clean energy, job creation," explains Corey Ramsden, vice president of Go Solar Programs at Solar United Neighbors, from his cavernous Dupont Circle office. "[It's about] relationships between people. People talk to their neighbors, increasing the network effect of solar. One neighbor goes solar. That builds on itself. Co-ops are not so much a formal governance structure as they are a way to build community."[20]

To better understand where we are and where we might go with solar, let's look at more of its history. Tech journalist Bob Johnstone offers an overview of 1970s off-grid origins as an alternative, hippie lifestyle to the multi-billion-dollar industry today. Both solar and wind were relatively insignificant contributors in their early years, with the U.S. government only beginning to report use statistics in 1984.[21] That year the two combined to deliver less than 0.0005 percent of US electricity. Only in 2008 did the United States first break the 1 percent threshold with solar, rising to merely 1.7 percent in 2020. Despite these meager percentages, the United States was the world's largest market for solar through the mid-1980s. President Jimmy Carter famously brought solar attention in June 1979 when he installed thirty-two panels on the White House's West Wing, although President Ronald Reagan removed them in 1986. And as oil prices fell that same year, making it even harder for solar to become competitive, the federal government slammed the door shut on our solar industry by eliminating tax credits for rooftop installation. Sales volume dropped 70 percent. Relinquishing our early lead in manufacturing, the United States was quickly overtaken by first Japan and Germany, then China. We've climbed back into the solar saddle the last few years, moving back ahead of Japan and Germany, but we continue to produce less than half what China does annually.

Americans believe we can do better.

Some 79 percent think our energy supply priority should be developing alternative sources of energy like solar, according to a 2020 Pew Research Center survey.[22] And at least one teenager in Florida agrees. In 2017, South Miami joined a half-dozen California cities in requiring solar on all new residential construction.[23] The city passed its 2017 law after high school student Delaney Reynolds wrote area mayors, imploring them to make the shift.[24] "We're down in South Florida where climate change and sea level rise are existential threats, so we're looking for every opportunity to promote renewable energy," noted then Mayor Philip Stoddard. "It's carbon reduction, plain and simple. We have a pledge for carbon neutrality. We support the

Paris Climate Agreement."[25] At only five to ten building permits a year it's a small step—but a step nonetheless. "It's not going to save the world by itself, but it's going to get people thinking about [solar]," as Mayor Stoddard says.

Thinking about solar begins with individuals igniting local political support. Research by Kathy Washienko, senior partner for climate strategies at Breakthrough, indicates people respond more enthusiastically to local action than national calls for patriotism, as President Carter attempted in his July 1979 malaise speech.[26] It's not installing solar panels on the White House roof that will motivate action. It's solar panels on local schools. That said, federal research and development support, as we saw in the 1980s, can be game-changers. And national initiatives like investment tax credits have been a driving force the last decade. Congress initially instituted a 30 percent tax credit (not deduction), which means reduction of tax liability. First scheduled to end in 2016, that credit was extended twice, most recently in a late December 2020 stimulus bill and continuing budget resolution. The credit was 26 percent for 2021 and 2022, dropping to 22 percent in 2023.[27] As John Klewin, chief operating officer of Clean Footprint, asserts, these "incentive programs make or break solar projects."[28]

So really, the record shows we need both local and national initiatives to foster a viable solar industry. And this means more than supporting solar as an emerging energy alternative. It means rethinking how we historically assist other forms of energy. National subsidies provided to the fossil fuel and nuclear industries, some experts argue, have the most deleterious effect, artificially propping up competition at below market prices. Estimates vary widely, but an often-cited global figure, using a 2017 International Monetary Fund (IMF) study that defines fossil fuel subsidies as unpriced pollution and greenhouse emissions, not merely direct payments from government, is $5.2 trillion.[29] The IMF says the U.S. portion in 2015 was $649 billion, notably more than our defense budget that year—and ten times federal education spending.[30] Other studies, such as one by Joseph Aldy at Resources for the Future (RFF), home in on only direct subsidies from federal tax credits, capital depreciation allowances, and financial treatments for royalties and passive losses. That means a much "lower" number of roughly $4.9 billion a year.[31]

Given our current fiscal and political environment, the most realistic solar initiatives remain regulatory support mechanisms at the state and local level. Renewable Portfolio Standards (RPS) are one example. Thirty states, plus the District of Columbia and three territories, have a RPS which mandates percentages of their electricity be generated from renewables each year. Another seven states have nonbinding renewable goals. States vary considerably in their specific targets, with most measuring requirements as a percentage of retail electric sales, although Iowa and Texas require specific amounts rather than percentages, and Kansas's requirement is a percentage of peak demand. About half the states target between 10 and 45 percent. But fourteen states (California, Colorado, Hawaii, Maine, Maryland, Massachusetts, Nevada, New Mexico, New Jersey, New York, Oregon, Vermont, Virginia, and Washington, as well as Washington, DC, Puerto Rico, and the Virgin Islands) have requirements of 50 percent or greater. All told, the National Conference of State Legislatures credits half the growth in U.S. renewable energy since the 2000s to RPSs.[32]

As noted earlier, net metering is another critical regulatory support. It's been instrumental in the United States for reasons already discussed, much more so than the widely and wildly praised Feed-in-Tariffs (FIT) in Germany. The German story centers around its 1990 feed-in law, which inserted market stability for investors by guaranteeing a twenty-year premium rate for those selling renewable electricity back to the utility. This meant homeowners not only broke even on their investments. They made profits as well. Grassroots activists were instrumental in shaping that first law, but a Red-Green governing coalition in 1998 also facilitated legislative success in 2000 along with revision of the Renewable Energies Law in 2004. The handful of FIT experiments in the United States, including one in Gainesville, Florida, were never able to replicate that success, though, due to a combination of natural gas competition and inordinate political power in the hands of utility companies.

That brings us back to the question of ownership. People get the biggest return on solar when it's on their own building, but many simply cannot afford that. Third party providers, through the aforementioned PPAs, open opportunities for more Americans by reducing restrictive upfront costs. Yet PPAs are prohibited in states like Florida because, as attorney George Cavros explains, "Utilities have oversized influence in Tallahassee. Their position is defined in statute and would require change in Florida law by legislature."[33] In short, utilities operate under an outdated business model, one whose priorities demand reconsideration. Utilities recover their costs by selling power—but they make their profit with large infrastructure investments such as power plants, power lines, and transformers. For those investments, they are granted, by law, a guaranteed rate of return. To date, utilities show no sign of changing that approach. "Utilities are not designed to move to new models; they never were," Zach Lyman, partner at Reluminati, an energy consultancy in Washington, DC, explains. "So they play an obstructionist role."[34]

It doesn't have to be that way. As attorney Cavros suggests, "Use performance metrics, instead [of infrastructure investment]. How well are they helping customers save energy? Go ahead and include reliability as well. We need a transformation to different metrics on how investors are rewarded. We've been using this method for over one hundred years now. It's time to switch to performance-based metrics where rates are tied to certain societal goals."[35] Becoming more energy-efficient is one of those societal goals. Creating more clean energy options with rooftop solar is another. But distributed energy means that homeowners are no longer passive consumers. They generate and store, even manage and economize, their own electricity. That means that consumers purchase less utility power, and utilities cannot justify as much infrastructure expansion. Their business model falls apart. For over a century now, utilities provided two services, the electricity itself and grid maintenance. Why not divorce the two and charge everyone for maintaining the grid, regardless of where their power comes from?

The future of solar is bright. Solar will be central in climate mitigation efforts, in decarbonizing our electric grid. It's an asset with climate adaptation as well, providing the additional power we'll need as climate change brings higher temperatures and more frequent disruptions to the traditional grid. Limitations remain, though, especially the issue of intermittency. Battery storage costs have not fallen as quickly as solar panel prices themselves and remain beyond the reach of many. "Storage is really the lever that will open wider access to distributive resources," states SUN's Ramsden. "How quickly we get there depends on how quickly prices come down."[36]

Even without batteries, though, solar offers stable pricing, a decided contrast to oil and gas and their fluctuation with global supply and demand, on average rising three to five percent each year. Rooftop solar also adds value for homeowners, on average $15,000, not to mention a boost in helping houses sell faster.[37] Debbie Dooley, who directed a two-year battle against solar restrictions in Georgia and helped found the Tea Party before joining the Florida SACE "Solar Choice" coalition discussed earlier, reiterates that economic rationale. "Who doesn't want to be able to have solar panels on their rooftops?" she asks. "Who doesn't want to become an entrepreneur—selling energy generated on their private property to their neighbors, and make a profit off of it?"[38]

The final game-changing characteristic of solar is its all-American emphasis on freedom and individual choice. Those resistant to climate change mitigation efforts often fear that their choice is being taken from them, that government is dictating how they live. Distributed solar addresses that concern as individual consumers take charge of their own electrical production. That's an approach that taps into conservative values and captures critical bipartisan support. "In poll after poll, 85 percent of Republicans support more solar," says Heaven Campbell, program associate with SUN FL. "It's about democracy. We really think that is the key here."[39] "We are intentionally and philosophically agnostic as to why people are attracted to solar," elaborates Ramsden, her colleague in Washington. "That makes it attractive in a bipartisan fashion. At the end of the day, it's about making sure there is more democratic access to producing electricity. Rooftop solar will be the cornerstone of a more equitable energy system."[40]

It's our responsibility to make sure we take advantage of the promise that provides. Join the next solar co-op starting in your area. If there's not one, start one. Two kids did. You can, too. But don't stop there. Be an advocate for your energy rights, your independence. Tell your friends and neighbors about the benefits you enjoy. Encourage them to explore those same opportunities. And take an active role in shaping state and local legislation that determines exactly how those opportunities play out.

10

LIVING WITH LESS

The summer sun never fully sets in Denali National Park. But every season has a counterweight, and, in Alaska, that translates into a mostly dark seven months of winter. For all its beauty, summer is a time of preparation for the harsh winter months ahead. Winter is always coming. And when that first frost inevitably arrives, Alaskan life has but three choices: hibernate, migrate, or tolerate. Alas, we've restricted ourselves to these same limited options with climate change. Over the last quarter-century, far too many have chosen option one by ignoring our unfolding climate crisis, dreamily wishing it away. Those around the world most immediately vulnerable, if they survive, have little choice but to employ option two and flee. And finally, those wealthy enough, while not yet sufficiently exposed to fully comprehend our climate crisis, simply tolerate it. Given warming projections, though, none of these will serve us through this century. We need a fourth option, one that mitigates climate change by reducing consumption.

I'm in the land of the midnight sun, a territory known for bountiful resources, gaining a better appreciation for what this entails. As the eight-hour Tundra Wilderness Tour winds down, our certified driver-naturalist, Sam, talks us through a meticulous recycling of our boxed dinner. Alaska's economic history revolves around extractive industries, from the fur trade and gold mining of another era to fishing and oil drilling today. But there is another form of consumption that forms the third leg in Alaska's contemporary economic triangle: tourism. Oil and fishing receive much deserved attention for their negative environmental impacts, but tourists like me cause damage, too. We on the Denali bus tour attempt to minimize that now. Recycling is a simple step, but its potential to shape how we consume holds a critical multiplier effect. The United States must become more aware in this regard. Our consumption patterns create an unsustainable carbon footprint. We live beyond our means, drawing down our natural capital account as if our children and grandchildren have no need for it. That's not sustainable. We must learn to live with less.

That's not exactly an inspiring rally cry. Americans historically pride ourselves on the opposite, our largesse. The closing of the American frontier in 1890, for example, sparked University of Wisconsin historian Frederick Jackson Turner to warn of irreparable impacts on American dynamism, innovation, and democratic ideals. The one saving grace then was plenty of open spaces remained across the continental states—and there was Alaska. Indeed, the impressive scale of our country, like the scale of Alaska, often distracts us from the finiteness of our land. With so much room to maneuver, it is understandable how some still see the solution to pollution as dilution, even when it comes to carbon dioxide and its greenhouse gas cousins.

Speaking of scale, when it comes to North America, they don't come any taller than the iconic snow-topped mountain peeking from between the fluffy clouds ahead. Athabaskans of interior Alaska appropriately named it Denali, meaning "the high one," and the centerpiece to this park stands an impressive

20,310 feet tall. The entire park totals six million acres, or almost 9,500 square miles. That's bigger than any national park in the lower forty-eight, but, as one more reminder of scale here, takes merely bronze in Alaska.

Only one road runs within the park, stretching parallel to the Alaska Range for ninety-two miles. Traversing picturesque low valleys and nerve-rackingly high mountain passes, the Denali Park Road provides another interesting window into consumption—having served as a focal point of debate about park facilities development. A classic wilderness versus development battle, the question was whether to serve more and more visitors or restrict access as much as possible. Park historian William E. Brown sums up Denali Park Road's significance succinctly as "the umbilical that would feed all other growth."[1] The paved portion ends after fifteen miles at Savage River and narrows significantly another fifteen miles later at Teklanika. This shows, according to wilderness proponents, evidence the National Park Service finally learned from its experiences in our lower forty-eight. During summer, roughly late May through early September, private vehicles may drive the first fifteen miles of this road – but are prohibited beyond that point. Tourists like me must use a shuttle bus system, and that public transit equates to less fuel consumption and a smaller carbon footprint.[2]

My kids are initially leery of this bus tour. The two older ones, as teenagers, are not particularly keen on spending eight captive hours with their parents, a bunch of strangers, and spotty cell phone service. But to their credit, they have become inquisitive travelers, and everyone is hooked after our first moose sighting. I'd like to say I adopted a bit of Athabaskan culture in shaping this segment of our adventure, acting like the raven, great trickster that he is, by applying deception and upturning of moral order to offer instruction on how children should behave. But in truth, my wife and I merely played our parental prerogative and gave them no choice.

Such paternalism is not going to work with the rest of our country, though. Instead of being told what to do, Americans need to discover it ourselves. We need an aha moment. For us to better live within our means, that aha moment must involve reconsideration of core values like consumption. This entails much more than symbolic changes. It involves much more than recycling a boxed lunch. American economist and sociologist Thorstein Veblen coined the phrase "conspicuous consumption" in his 1899 *The Theory of the Leisure Class* to describe purchasing as a medium to flaunt wealth. Veblen viewed such socially driven consumption as negative, and it often is.[3] That's magnified more when one considers, of all the materials in our consumer economy, only one percent remain in use six months after sale, as Annie Leonard explains. We are liquidating earth's natural assets to fuel this consumption—but not even using those materials for half a year.[4] That should raise eyebrows, especially after hearing that our current consumption levels really require one and a half Earths.[5]

Alas, some still see no problem here. They believe continued progress will be met by advances in technology. With no limits to growth, their faith rests in infinite progress. "The main fuel to speed the world's progress is our stock of knowledge; the brakes are our lack of imagination," argued cornucopians like the late economics and business professor Julian Simon.[6] The more population the better, their thinking goes. More people mean more inventions, more wealth. But we now understand that's not true. Between 60 and 80 percent of our impact on the planet comes from simple household consumption.[7] To quote President Jimmy Carter years ago, "Too many of us now tend to worship self-indulgence and consumption. Human identity is no longer defined by what one does, but by what one owns."[8]

"Inertia is our worst enemy," elaborates Lester Brown, founder of DC-based environmental research organizations Worldwatch Institute and Earth Policy Institute. "[We require a] Copernican-scale shift in economic thinking. Planned obsolescence, the throwaway products seized on enthusiastically in the United States after World War II as a way of promoting economic growth and employment [drive this problem]."[9] To move beyond this throwaway society, as Ann Francis, former sustainability coordinator at Rollins College, suggests, we must better consider the repercussions of our consumer habits.

It turns out that direct experience shapes how we make such recalibrations. "Students learn more when it comes to sustainability and environmental issues when they do a hands-on project," Francis explains. "When they see those things, I think they go 'hmmm.' I've seen a lot of students become more interested, and didn't even know they were interested, until they saw it themselves."[10] Let's take another look at recycling in that context. Contamination was a huge problem in the program Francis helped build at Rollins. It was a single-stream system: anything and everything recyclable went into the bins. For years, that seemed to work, but then a student honors thesis demonstrated otherwise. Food and beverages weren't contaminating the recycling efforts. The blue plastic liners used to collect the recycling were. In single-stream recycling, everything is tossed onto rollers and separated into different categories, but the plastic bags that collected the trash were clogging up those rollers. Workers were spending hours cutting plastic bags off rollers because they stopped the machines. In time, as soon as they saw a plastic bag, they just took it to the landfill. Those materials were never recycled. When Rollins discovered this, they ended single-stream recycling and concentrated on recycling paper products and cardboard along with metal.

Plastic, though, remained problematic. Recognizing recycling is the last stage of the age-old mantra reduce, reuse, and recycle, Francis had an idea. To improve the campus carbon footprint, reducing and reusing should come first. With that, she redoubled efforts to encourage reusable bottles, and some fifty hydration stations facilitating that now dot the eighty-acre campus. Another important lesson here is that initial failure doesn't always translate into a final negative result. That's something we must also come to better grips with. Failure need not be permanent. It can teach valuable lessons in how not to do something—and how to do it better. The simple fact is we don't live in an ideal world. And students' experience with recycling complications was a good lesson in that. That experience also emphasizes that how we consume matters. Rollins students didn't stop drinking water. They changed how they drank it.

Packaging constitutes as much as one-third of solid waste generated by households. Much of this goes to landfills, where we find the largest human-related source of methane in the United States, a quarter of all our emissions. Methane is a powerful greenhouse gas, with a hundred-year global warming potential twenty-five times that of carbon dioxide (and eighty-four times as potent in a twenty-year period).[11]

Not all trash finds its way into landfills, though. Take plastic as an example. It accounts for 10 percent of all fossil fuel burned on the planet because it is made from oil. Most Americans don't even know that. They don't know one major by-product of any oil refinery is something called "petrochemical feedstock," the raw material from which plastic is made.[12] Plastic offers many advantages. Sanitary packaging reduces spillage and extends the shelf life of meat and vegetables from days to weeks. Plastic replaced items previously made from glass, metal, paper, and cotton, because it is lighter and cheaper to produce. Plastic is also easily sterilized, leading to medical popularity in the form of containers, disposable syringes, tubing, and even artificial corneas. And cars, trucks, and planes are considerably more fuel efficient thanks to plastic, with replacement of heavy metal parts by lighter molded plastics beginning in the 1980s. None of this was possible before World War II, yet global production today is more than three hundred million metric tons per year, about equal to the total weight of all persons on Earth.

That's not necessarily something to brag about, particularly as a third of plastic consumed in the United States is nothing more than packaging. Excessive packaging of fruit and vegetables is problematic, but bags and bottles are the biggest bad guys. A Swedish engineer designed the polyethylene shopping bag in 1965, and they quickly began to replace cloth bags in Europe, controlling 80 percent of the market by 1979. The largest U.S. supermarket chains converted to plastic in 1982. By the 1990s, they had spread worldwide. The average American uses between three hundred and seven hundred plastic bags each year. That translates, just in the United States, to 12 million barrels of oil annually.[13]

Plastic bottles are nearly as ubiquitous—and even more problematic. Nearly 15 billion gallons of bottled water were sold in the United States in 2019, and that number continues to rise.[14] To package water, the United States alone uses 17 million barrels of oil. Add the energy required to refrigerate and transport it, and our bottled water industry consumes up to 50 million barrels of oil per year. That's equal to 13 percent of U.S. oil imports from Saudi Arabia.[15]

On average, someone within the wealthy thirty-seven states of the Organization for Economic Cooperation and Development (OECD) discards at least their body weight in plastic each year, with the United States leading the way.[16] Yet globally, less than 9 percent of plastic is recycled,[17] disposed instead within landfills and poorer countries where it is cheaper to dump. China was a leading recipient of discarded plastic for years but stopped accepting waste from the West in late 2017. Indonesia, Vietnam, and the Philippines now fill this void.

But it's not just developing countries that amass our unwanted plastic trash. The oceans also suffer. Almost 10 percent of what we throw away ends up in them. One of the five major offshore accumulation zones is something called the Great Pacific Garbage Patch. Twice the size of Texas, straddling Pacific waters between California and Hawaii, it's a sea of plastic scraps, from bottle tops and beverage cups to broken toys and those ubiquitous plastic bottles. And then there are microplastics, plastic fragments ranging from five to ten nanometers. A good chunk of this comes from polyester or nylon, the most used materials for clothing today, as they shed every time they're washed. A single machine load releases up to 700,000 fibers, according to research from the UK's University of Plymouth. Roughly one-third the plastic in our oceans comes from washing our clothes.[18]

There is another image of plastic you might have seen, one that is difficult to forget. It's the plastic found blocking airways and stomachs within birds, turtles, and dolphins that mistake it for food. Ninety percent of seabirds likely have plastic in their stomachs. A quarter of the fish sold in California contains it. Most fish tested in the Great Lakes and 73 percent of fish surveyed in the northwest Atlantic do, too. Ultimately, this plastic ends up in our own food system. If you eat fish, there is a 99 percent chance you ingested microplastics.[19] But you can ingest plastic even without eating fish. Bottled water often sits in its plastic container for months, off-gassing chemicals and leaching microplastics.

Plastic consumption is not only a resource and climate change crisis, then, but a bird, fish, turtle, and human health crisis as well. Cleanup attempts are a Sisyphean task with new waves of waste, at least those not ingested or caught in one of those five major oceanic gyres, floating ashore each tide. More than sixty countries have taken proactive action, issuing bans and levies on single-use plastic. We, in the United States, have not. The Break Free from Plastic Pollution Act introduced in Congress in early 2020 would reverse that poor record. Seeking to ban many single-use plastics nationally, that legislation would also prohibit plastic waste from being exported to developing countries and mandate packaging producers finance waste and recycling programs.

While the future of this federal legislation remains in doubt, local initiatives again fill the void. San Francisco started phasing out plastic bottles in 2007 and banned single-use plastic bags in 2012. Los Angeles followed with legislation restricting plastic bags in 2013. The entire state of California banned major retailer usage in 2014. More recently, Los Angeles voted in 2019 to phase out single-use water bottles at all city events and in all city-owned buildings, including Los Angeles International Airport and the LA Coliseum.[20] This isn't just a left coast issue. We might not be as trendy as our California cousins, but the East Coast version of Orange County in Orlando, Florida, banned use of plastic straws, bags, and utensils on all city property in 2019. If we can do it, you can too. Talk to your friends and neighbors. Talk to people at your children's schools and your place of worship. Then petition your local government.

It's not merely the packaging of our goods that we need to address. How goods are produced and where they come from matters, too. Nothing illustrates this better than food. Our modern food system comes with substantial social and environmental costs, largely due to industrialization of agriculture over the latter half of the twentieth century. More centralized food production, with fewer American farmers today than during the Civil War, means food travels further to reach our plate. For a healthier climate, as bestselling author Jonathan Foer contends, the future of farming needs to better resemble the past.[21] Twenty-five percent of US greenhouse gas (GHG) emissions are traced to food, with agriculture production alone contributing ten percent.[22] Global food production also drives much deforestation, which accounts for one-fifth of global warming emissions.[23] Tropical deforestation, driven by soybean and beef production as well as lumber extraction, is the second biggest cause of global climate change, trailing only burning of fossil fuels. Even if we abandoned fossil fuels tomorrow, our current food system would produce emissions pushing us past Paris targets.

The slow food movement suggests an alternative route. Started in Italy in 1986, it prioritizes traditional cuisine along with plants and livestock local to a region. This emphasis on local reduces emissions. Shipping goods, for example, creates about half today's transportation emissions, although admittedly not all of that is from food.[24] While it matters if this is with carbon-intensive trucking or less polluting ships and trains, shipping less, by shifting to more local production, also has profound political repercussions. Some 75 cents of every dollar spent on supermarket food is consumed by advertising, packaging, long-distance transport, and storage, whereas, at farmer's markets, 95 cents on the dollar goes directly to the farmer growing the food.[25] Food miles, the number of miles a product must travel to reach its consumer, should play a larger role in decisions about food consumption.

Buying local saves energy, reduces our carbon footprint, and builds stronger communities.

Less meat is also part of this equation. For one, animal agriculture is the leading cause of deforestation. For another, livestock are a leading source of greenhouse gases like methane and nitrous oxide (from urine, manure, and fertilizers utilized growing feed crops). On top of all that, it takes eleven times as much fossil fuel to raise a pound of animal protein as a pound of plant protein.[26] If cows were a country, they would rank third in greenhouse gas emissions after China and United States, according to the United Nations Framework Convention on Climate Change (UNFCCC).

It's not just planetary health that should worry us. Our individual health is at risk, too. Americans consume twice the recommended intake of protein, most of that from meat. People who eat diets high in animal protein are four times more likely to die of cancer. They are also more likely to suffer strokes and heart disease. Livestock use 70 percent of antibiotics globally, thus weakening antibiotic effectiveness with human diseases. Foer recommends that Americans skip meat until dinner, suggesting true climate-friendly consumption would amount to 90 percent less beef and 60 percent less dairy on dining tables.[27]

Personal transportation is another key sector where less is needed. Long flights to Alaska, like mine, should be the exception rather than the rule—and must inspire cuts in carbon upon returning home. Furthermore, more common, short-haul flights under 300-odd miles are among the worst polluters when it comes to air travel owing to the energy required to rise during takeoffs as well as return with landings. So reducing the frequency with which we take shorter flights should be on our individual agendas as well. More accurate pricing of these flights should consider greenhouse gas emissions by charging less for direct flights than those with connections.

Still, even as air transport emissions grow and deserve our attention, three-quarters of global transport emissions come from road travel, most from passenger vehicles.[28] The more immediate and more local focus with transportation emissions, then, belongs on the ground, namely the car culture that dominates daily American lives.[29] More accurately priced gasoline, accounting for its externalities, would force Americans to recalculate our driving interests, reducing consumption and mitigating climate change by nudging people toward public transit or at least more fuel-efficient personal vehicles.[30] Alas, a gasoline tax is regressive, with its impact felt among lower income populations disproportionately. That is why public transit, as we will explore in chapter 14, is such a critical component to reducing greenhouse gas emissions equitably.

Biking is another alterative. "We can't continue to accommodate a lot of the growth with cars," explains Polly Trottenberg, former New York City transportation commissioner. "We need to turn to the most efficient modes, that is, transit, cycling and walking. Our street capacity is fixed."[31] New York City developed the first bike path in the country in 1894 along Ocean Parkway in Brooklyn. Today, it's one of more than one hundred American cities with protected lanes as buffers between bicycles and cars. Portland is another. The City of Roses has 370 miles of bike routes, along with "neighborhood greenways" that encourage biking by lowering auto speed limits to 20 mph on neighborhoods streets. Biking not only reduces our carbon footprint. It saves money and provides Americans with much needed exercise. "The city has gotten a lot more crowded, and the trains have gotten a lot more expensive," says Jace Rivera, a former construction worker who enjoyed riding a bike to work so much he changed careers to become a bike messenger. "By biking, you spare yourself the crowds, you save a lot of money, and you can go to work on time."[32]

A final arena of unsustainable consumption involves residential construction. Construction cement requires huge amounts of energy and accounts for about 4 percent of global carbon dioxide from fuel use and industrial sources. Reducing our carbon footprint means investing in alternative materials and design as highlighted soon in chapter 12. But it also means living smaller. That entails reversing a trend where the average American single-family house size doubled the last half of the twentieth century, even as occupant numbers shrank. Living with less also means living closer, moving away from the suburban sprawl model dominating places like Central Florida, where I live. The Healthy Community Initiative of Greater Orlando defines sprawl as: "unnecessary land consumption—repetitive one-story commercial buildings surrounded by acres of parking, and a lack of public spaces and common centers. It is poorly managed growth with disregard for a sustainable balance of nature, economy, society, and well-being." As late environmental writer Bill Belleville lamented, "Sprawl destroys green space, increases traffic, crowds schools, and drives up taxes."[33]

Political ramifications here are far-reaching. "Sprawl isolates people in their own homes . . . turning America into a society of strangers," as Douglas E. Morris asserts in *It's a Sprawl World After All*. "Because no one knows anyone else, it has helped to create a culture of incivility. People realize that if they're never going to see anybody again, they can be rude and uncivil."[34] Florida is infamous in this regard. My state is losing its connectedness, its link between people and the land. To add insult to injury, this sprawl is subsidized by taxpayers as we assume the added expenses it incurs. According to the Florida Hometown

Democracy coalition, for every $100 in tax revenue new development generates, it costs $130 in government services such as roads, schools, law enforcement, and sewage treatment.[35]

A discussion over coffee one November morning with Paul Owens, President of 1,000 Friends of Florida, fleshes this out further. Owens previously served as opinion editor for the *Orlando Sentinel*, writing extensively on growth management, environment, and quality of life issues facing the state. He's kept that same passion in his new role, spreading his nonprofit's message of bipartisan advocacy for building better communities, saving special places, and fighting sprawl. According to Owens, we are losing ten acres of rural and natural land an hour. In my pigskin-crazed state, that equates to seven football fields every sixty minutes. But experts still struggle how to tell this story. Owens points out the term growth management doesn't exactly turn heads and is "hard to put on a bumper sticker." He is also keen to emphasize his group is not opposed to growth altogether, merely poorly managed growth. Yet, like Rachel Cason's 1960s fight against indiscriminate use of pesticides, that message is easily twisted into something it's not. Owens and a growing number of Floridians support responsible growth. As he explains, the sunshine state should better "focus our growth in areas where it makes sense, in urban areas that have infrastructure in place to handle it. And then we can start investing in transportation improvements in those areas—and affordable housing."[36]

Like climate change itself, this is not a red or blue issue, but it, mistakenly, often becomes one. Polarization positions itself as part of the problem. "If you are advocating for environmental stewardship, then it is assumed by a certain share of your audience that you are going to be checking the boxes on a whole range of issues that are not directly related," contends Owens. "I think that promotes suspicion and resistance among a certain share of the folks that are listening to you."[37] That's counterproductive. Climate change should be something that unites us, not divides us. There are at least as many conservative reasons as liberal to support efforts to mitigate it. "Climate change is used consistently in our messaging," Owens continues. "We don't censor that. We try to emphasize this is not part of any party's political agenda. It's in Florida's best interest, and it's in their best interest as Floridians, to want policies that directly address this threat to their way of life, to the future for their kids and their grandkids, and again to their property values. We are scrupulously nonpartisan. We are willing to work with any politician in Florida who we believe will promote responsible future-oriented policies."[38]

Reducing consumption levels is not just an American problem. China is central to our predicament. While there is much to critique in the East, though, there are lessons to learn there as well. The Chinese State Council announced in late 2007, for example, that all supermarkets and department stores would no longer distribute free plastic bags beginning June 1, 2008. That took half a year. In contrast, the United States started a twenty-two-year process of removing lead from gasoline in 1973. I realize this isn't exactly an apples-to-apples comparison. Leaded gasoline was more difficult for China to phase out as well. Its government committed to lead-free in 1998, implementing it in 1999. In under two years, in 2000, China was lead-free. Of course, China is not a democratic society, and that allows shortcuts we wouldn't want to emulate. But twenty-two years is too long, particularly when compared to less than two. And we haven't exactly been winning awards for our democratic processes lately.

The United States needs to get its act together—and that starts with you. Individual actions are admittedly small steps. But small steps can lead to large results. Rake instead of blowing leaves. Bring your own mug to work instead of relying on disposable cups. Forsake bottled water. Invest in cloth shopping bags instead of petroleum-based plastic ones. And yes, eat less meat. These may all seem relatively insignificant at first blush, but breaking our climate crisis into smaller, more manageable problems is an important first step.

Take the Y2K problem at the turn of the century as an example. This troublesome computer bug lay dormant within nearly all computers. It was a highly technical problem, one not immediately obvious to most, with some even questioning whether it existed. Sound familiar? Oh, and Y2K was prohibitively expensive as well. The US spent $100 billion.[39] Worldwide we spent $580 billion.[40] Yes, climate change is even more expensive. The United Nations estimates that $300 billion will be needed over the next twenty years, and in the long term we will spend trillions instead of billions. Yet, what drove the Y2K action is still informative. "Y2K shows that the way problems are portrayed is crucial to how solutions are approached," two Australian researchers, John Phillimore and Aidan Davison, argue. "Small, discrete problems are easier to understand than 'slow-burn,' incremental ones. Providing specific examples of things that might go wrong is more effective than general warnings . . . this might be particularly pertinent to debates on global warming."[41]

This seems simple. But it's not easy. Most Americans prefer upstream solutions targeting producers rather than themselves as consumers. Researchers at Georgia State University found that emphasizing personal behavior can reduce people's willingness to address climate change. "Messages about policies that would affect others, such as taxes on industry and business or on carbon emitters," they write, "are more palatable and do not result in such a negative response."[42] That said, past examples of successful personal sacrifice do exist. During World War II, Congress lowered the minimum taxable income while reducing personal exemptions and deductions. Only 10 percent of American workers paid federal income tax before we entered the war in 1940, but, by 1944, nearly all did. You've heard of war bonds, victory gardens, and Rosie the Riveter. The traditional third rail of American politics, taxes, played a role also.

The main difference today rests with who our enemy is. In World War II it was the fascist states of Germany and Japan. That's not the case with climate change. "We have met the enemy, and he is us," as cartoonist Walt Kelly famously parodied in his 1970 anti-pollution Earth Day poster.[43] This means addressing not just the supply side of consumption in the form of multinational corporations, big ag, developers, and Madison Avenue. It means targeting the demand side, too. It means, like democracy, the responsibility ultimately rests with us. "Our obsession with consumption and individual rights, to the neglect of collective rights," as academic David Orr writes, is unsustainable.[44] We would be better served returning to the more traditional American values Orr identifies like thrift, frugality, and neighborliness. It's really that simple. "There are two ways to get enough," as the English writer and philosopher G. K. Chesterton once stated. "One is to continue to accumulate more and more. The other is to desire less."

That's on all of us. How and what we consume matters. Small steps add up, from eating local, consuming less meat, and reducing food waste to driving less and biking more, and avoiding plastic packaging, bags, and bottles whenever possible. But individual choices like these will not be enough—unless they force systemic changes. Borrowing from climate scientist Katharine Hayhoe, successfully scaling up these actions depends upon us all doing a better job of talking about climate change, both its causes and its solutions.[45] We all care about this land. Regardless of how stark our differences in ideology may be, common ground remains centered around that shared value. Solving climate change, then, requires rethinking what, how, and why we consume. It means recognizing that sometimes less really is more.

11

THE WINDS ARE CHANGING

Even small towns go big in the Lone Star state. The modest West Texas town of Sweetwater, population 11,415, for example, proudly bills itself as annual host to the World's Largest Rattlesnake roundup. And every March, on average, locals capture four thousand pounds of the venomous reptiles. The practice dates to 1958, when farmers and ranchers, attempting to protect their livestock and pets, sought to eradicate their western diamondback population. Snakes are still the main event today, weighed, sexed (identified as male or female), milked, killed, and skinned. There's also a cookoff, gun and knife show, carnival, and flea market. Together, that eclectic combination attracts some 25,000 visitors who inject roughly $8.3 million into the economy of this Nolan County seat. Alas, I missed those festivities by a couple months and am here forty-one miles west of Abilene learning more about another arena in which Sweetwater decided to go big: wind. Texas produces more wind energy than any other state in our country. In fact, it produces more wind energy than the next three states combined—with tiny Sweetwater straddling the center of this production.

That's a big deal for what's still an oil and gas state. Fossil fuels remain politically prominent, shaping state culture and dictating finances. But wind is a major player now, too. "Texas gets about 20 percent of its electricity from wind alone," says Mark Z. Jacobson, a professor of civil and environmental engineering at Stanford University and senior fellow at the Stanford Woods Institute for the Environment, as well as author of a study looking at the future of smart grids.[1] And this West Texas horizon holds much more. If Texas were a country, it would rank fifth in the world for wind power capacity.

Farmers and ranchers cursed this wind for decades. It dried out land, killing crops regularly. Times have changed. Most now see wind as a blessing, bringing $5,000 to $15,000 per turbine a year, depending on tower size. As expected in Texas, where everything is bigger, those towers tend toward the tall. Typically mounted two hundred feet off the ground, approximately twenty stories up, each turbine blade is the size of an airplane wing. Blade to blade, that's longer than a football field. And there are hundreds of them in this area alone. All total, the Lone Star state holds 14,720 wind turbines, according to the U.S. Office of Energy Efficiency and Renewable Energy.

Small as it is, Sweetwater was a historical hub for cotton, oil, and cattle. Rail shipping continues for several manufacturing interests. And Interstate 20 shoots 34,000 vehicles by the town every twenty-four hours, so it's not exactly isolated. But Sweetwater would still be figuratively in the middle of nowhere if not for wind. Economically, wind has been a game-changer, saving numerous farms and ranches. Most of the year, this land sees little rain, totaling only twenty-three inches annually. Then, on occasion, there are bursts of too much of it. Eight or nine inches fell the last two and a half weeks, for instance, flooding acres of cotton fields before me. Before wind, that might have sounded the death knell for a family farm. "Many farms couldn't make their tax payments when a flood or drought hit in the past," explains Ken Becker, executive director at Sweetwater Enterprise for Economic Development (SEED), while driving around Noland County for a couple hours one breezy May afternoon. "Wind allows them to keep their land in the family for generation to generation."[2]

When margins are thin in this climate, moreover, fewer opportunities to try something different exist. "Because of wind, farmers and ranchers around Sweetwater have been able to diversify," continues Becker. "They are even experimenting with no till cotton so as not to lose nutrients and moisture. The fact they had wind income allowed them to take some gambles not everyone else is willing to take." Even landowners without wind turbines on their property have figured out how to benefit from it. Miesha Adames was the first to sign on with the 2005 Competitive Renewable Energy Zone (CREZ) expansion in the Nolan County area, selling ten of her acres to allow construction of three sets of transmission lines, while keeping her remaining 1,165 acres for hay, wheat, and cattle. "Without that extra income people like me wouldn't be able to afford their property," says Adames.[3]

Then there are the wind jobs themselves. Wind turbine technician positions are among the fastest-growing occupations within the entire American economy, followed by nurse practitioners and solar photovoltaic installers, according to the U.S. Bureau of Labor Statistics.[4] In Texas, more than 25,000 work within the wind industry, many coming from rural communities devastated by declines in agriculture and manufacturing over the last several decades. That's a big win for local folks' pocketbooks. It's also good for the tax base and local communities overall. "Farmers and ranchers get the income from the land, the communities get a bigger tax base, and the cities get a cheap, stable energy source," asserts Susan Sloan, formerly vice president for state affairs at the recently renamed American Clean Power Association in Washington, DC.[5]

Sweetwater is doing its part to train this workforce, too. Three miles west of its historic downtown sits Texas State Technical College. Set on the site of a former airbase, the school offers nursing, welding, diesel as well as auto mechanic, and electromechanical training courses. Wind also became part of the curriculum over a dozen years ago. It didn't take long to blossom. "Right now, wind is our number one career," states Billie Jones, statewide department chair for Wind Energy Technology with Texas State Technical College. "Basically, my students have a job before they graduate."[6] Those students come from all walks of life, from high schoolers with dual enrollment to those over fifty seeking a new career. Jones, a former GE turbine tech for eight years, understands exactly why. "I loved my job as a turbine tech," she says. "The pay is good, and there's an addiction to climbing the turbine, being able to sit in the clouds."

Technically, wind does not blow. Rather, it's drawn from areas of higher to lower pressure. Temperature differences from the Earth's surface up to the stratosphere cause these pressure differences and generate air currents we know as wind. When it comes to harnessing this energy, strong and steady winds like those on the mesas here in West Texas are best. Winds whipping off large buildings or tall rocks are not as desirable, producing less power while submitting equipment to more wear and tear. The two main types of wind are onshore and offshore. Offshore turbines tend to be even larger in scale than those onshore. They also carry a larger ecological footprint, especially as they typically require foundations drilled into the ocean floor as well as installation of underwater transmission lines. That said, offshore turbines carry environmental benefits as well, acting as artificial reefs for marine organisms.[7] And less invasive floating platforms might be cost-competitive by the late 2020s.[8]

For the moment, though, most wind turbine attention in the United States, where five of the ten largest wind farms in the world operate, is onshore; only one commercial offshore wind farm currently operates in the United States, with five turbines off the coast of Block Island, Rhode Island.[9] Our Great Plains, specifically North Dakota, Kansas, and Texas, are the Saudi Arabia of wind power. Those three states, according to a national wind resource inventory by the U.S. Department Energy decades ago, could provide enough electricity for our entire country, coast to coast. We aren't there yet, of course, but forty states, including Hawaii and Alaska, already use utility-scale wind farms for electricity generation. And those states producing the most wind energy are a diverse mix, with Texas leading the way, followed by Iowa, California, Oklahoma, and Illinois.[10] The Lone Star State also boasts six of our country's ten largest wind farms, according to the U.S. Energy Information Association (EIA). The biggest of all, the Roscoe Wind Farm, neighbors Sweetwater, with 627 turbines spread across four counties. It produces enough electricity to power 265,000 homes.[11]

All this begs the question: Why Texas?

There are at least half a dozen reasons. For one, Texas holds the ideal physical terrain with wide swaths of unobstructed space. That means, in highly technical terms, "Texas has a boatload of wind," as Chrissy Mann of the Sierra Club's Lone Star chapter articulates.[12] Texas journalists Kate Galbraith and Asher Price make a similar point, using state livestock to illustrate the power of wind in these parts. Because wind is so strong, "Cattle could die, drown really from inhaling snow that blew horizontally, with tremendous force, during a blizzard," they write in their history *The Great Texas Wind Rush*.[13] But while the state sits securely within the wind belt, straight up the middle of the country to Canada, the states of Oklahoma, Kansas, Nebraska, and the Dakotas, as well as portions of Minnesota, Iowa, Wyoming, and Montana, all sit there as well. Wind potential alone cannot explain Texas' lead position.

Deeper analysis points to two pieces of legislation passed under two different Republican governors, neither considered a friend of the environment or opponent of fossil fuels. First, in 1999, Governor George W. Bush signed an electricity deregulation law that split apart electric utility companies into power generating companies, transmission and distribution utilities, and companies that sold power to customers. Senate Bill 7 also established a statewide requirement on renewable energy. Then, in 2005, Governor Rick Perry signed into law the $7 billion Competitive Renewable Energy Zone (CREZ) initiative that rancher Miesha Adames, as noted earlier, utilized. That legislation added 2,400 miles of transmission lines connecting the windy plains of West Texas and the Panhandle to population centers along the Interstate 35 corridor, from Dallas/Fort Worth to Austin and San Antonio. "It's really the expansion of Texas's transmission lines that explain Texas's success," argues Sarah Mills, an engineer and development expert at the University of Michigan who studies wind energy in rural areas.[14]

That's partly true. But it's not just the CREZ lines that spurred investments. Texas blew past California to become our wind-power leader in 2006, well before these new CREZ lines started coming online in 2013. Decades earlier, dating back to the 1950s, oil and gas primed Texas for wind with their own transmission line construction. "The wealth beneath the land had caused people to build power plants out in middle of desert, and with those gas plants came transmission lines to move power from remote mesas to big cities, like Dallas or Austin or Houston," contend journalists Galbraith and Price. "Wind turbines in the middle of nowhere could piggyback on those power lines."[15] Sweetwater's Ken Becker agrees. "One of the reasons why wind came here is the power transmission lines were already built," he asserts, as we drive underneath a set of them. "They built them in the 1950s from Dallas to Midland and Odessa because of the oil and gas big boom, thinking they were going to need a lot of power to run their pumpjacks. They didn't realize it wouldn't take that much power."[16]

Perhaps the most unique piece to the Texas puzzle, though, is how the state was able to build. The continental United States is divided into three electrical grids: East, West, and Texas. That means that Texas, at least 75 percent of it, owns its transmission lines and is not subject to the commerce clause of the Constitution, which gives the federal government power to regulate interstate commerce.[17] Texas can act on its own.

Another factor unique to Texas, particularly compared to states farther west, is that over 90 percent of the state is privately owned. When it joined the union in 1845, after nine years as a sovereign nation, private citizens already held title to most of the land. Conversely, in Arizona, New Mexico, and California, 30 to 40 percent of land is federally owned, meaning energy developers may face extensive federal regulatory hurdles. That's not the case in Texas. "Texas doesn't have to receive federal permits, so it's able to move more quickly on infrastructure," said Robert Stavins, a professor of energy and economic development at the Harvard Kennedy School.[18]

Connected to that historical quirk is a tradition of rural self-sufficiency, a just-do-it attitude that borders on anti-environmentalism owing to a state economy built on extracting resources. The mix of large rural areas with minimal zoning restrictions further contributes to a culture limiting government intervention in private property. "There aren't a ton of rules about where things get put," explains Mills. "And there are lots of landowners with large swaths of land who can place multiple turbines."[19] Sometimes, like in the case of wind, this can be an environmental plus. "In Texas, because we don't care about the environment, we're actually able to do good things for the environment," asserts Michael Webber, an energy expert at University of Texas at Austin.[20]

That said, a final condition shaping Texas's wind leadership is strong grassroots support, thanks to the fact that there is money to be made. "The political movement came from the West Texas power base, with farmers and ranchers and county judges out there," says Roger Duncan, a board member of the Alliance to Save Energy and former general manager at Austin Energy. "That is the power base that went to the Texas legislature and said, 'We need incentives and goals for wind, because it can make money.'"[21] That political base grew further as counties and school districts sought to cash in on the increase in property tax revenues that wind turbines sited on farms and ranches would bring.

Humans have harnessed wind for transportation over thousands of years, sending goods up rivers and over oceans. Egyptians employed wind to navigate the Nile as early as 3100BCE, while centuries later Greeks and Romans built their respective Mediterranean Sea expansions around wind and sail. Historians trace the first devices for grinding grain and drawing water to tenth-century Persia. Returning crusaders likely introduced these practices to Europe in the twelfth century, and between the fourteenth and nineteenth centuries approximately a quarter of the continent's energy needs came from wind.[22] In between, during the seventeenth century, the Dutch built thousands of windmills to pump out water for their massive land reclamation projects. And American engineer Charles Brush built the first windmill to generate electricity in 1888. With little use for electricity before the twentieth century, though, wind electric technology only began to progress after World War I thanks to wartime experience with aircraft propellers.

Experimentation continued in the years that followed, with the modern wind energy industry emerging in California in the early 1980s. Spurred by oil price hikes in 1973 and 1979 as well as a young Governor Jerry Brown, a mix of environmental activism and entrepreneurialism shaped this initial growth burst. Technology advanced again in the mid-1990s as manufacturers began building variable-speed turbines to capture more energy while reducing mechanical stress.[23] Reliability and power output improved—and that trajectory continued over the next two decades. According to the Department of Energy, the average capacity factor[24] of wind projects installed from 2004 to 2011 averaged about 30.8 percent. Between 2014 and 2017 it was 41.9 percent.[25] Perhaps most important, though, overall wind costs dropped dramatically during the last decade, falling 70 percent and positioning wind as the most affordable new electricity source for most of the country.[26] Wind electricity priced out at $.55 per kWh in 1980, for example, but under $.03 today, according to the Department of Energy's Office of Energy Efficiency and Renewable Energy. That's huge, because public support evaporates without clear financial benefits.

Wind turbines exaggerated reputation as a bird and bat killer also threatens public support. One of the earliest wind farms in the United States, California's Altamont Pass, for example, straddles a major migratory path and claims between 160 to 400 birds a year, according to an early study. "But birds fly into houses much more than they fly into turbines," notes Marianne Rodgers, scientific director for the Wind Energy Institute of Canada.[27]

Digging deeper, one finds turbines responsible for less than 0.01 percent of annual avian mortalities. Cats are considerably more dangerous, claiming 10 percent of bird kills, whereas power lines are responsible for 13 percent and buildings a whooping 58 percent.[28]

Ice throw from rotating blades during winter months is another complaint. The damage done to roads from shipping heavy equipment and material to construct them is, too. But one of the more intriguing criticisms of wind turbines, one with questionable merit, focuses upon health effects. The charge here is their strobe effect, the flickers between shadow and light as blades rotate, causes physiological effects like heart attacks or vertigo. Some say they are too noisy and complain their low frequency emissions and infrasound combine with vibrations to cause something called Wind Turbine Syndrome.

No evidence on this exists, but turbines do make noise. Standing beneath, one draws comparison to ocean sounds, with waves of wind breaking rhythmically as turbine blades rotate. To some, like me on many an occasion, that can be a peaceful way to fall asleep. Then again, I also sympathize with someone whose sleep may be disrupted by the combination of vibration and low frequency noise emissions, particularly since my neighbor's pool pump has been on the fritz the last couple of evenings. But the fact that my wife is having no such sleep difficulty gives me pause. Perhaps what I hear, maybe like some that complain about wind turbines, is merely tinnitus, the perception of noise or ringing in your ears. The phenomenon is relatively common, affecting 15 to 20 percent of our population, and is a symptom of an underlying condition like age-related hearing loss, ear injury, or circulatory system disorder.

Beyond these complaints, the most powerful wind opposition to date centers on aesthetics. Wind turbines change how a landscape looks, altering previously scenic and serene settings. As such, surrounding property owners, fearing quality of life and property value losses, exhibit a classic NIMBY (not in my backyard) response. Even more than wind farms, though, transmission lines spark understandable opposition. To many, wind turbines exhibit calm and elegance while "transmission lines are simply wires stretching across the horizon . . . cut[ting] through the landscape like a long, thin scar . . . for miles," in the words of journalists Galbraith and Price.[29]

These controversies aside, consensus regarding other limitations with wind exists. For one, wind turbines eat up land, requiring six times more acreage per watt produced than solar. Construction, requiring concrete, is not emission-free. There's also the problem of disposing decommissioned blades as turbines are updated. But the biggest disadvantage of all is intermittency. Wind is even less predictable than sunshine. The strongest winds are in springtime, but our grid needs power most in late summer afternoons, when West Texas breezes frequently come to a standstill. There is also the problem of too much wind. When wind speeds reach an upper safety limit of 50 to 55 mph, turbines must shut down or risk destruction. And finally, while wind complements solar, tending to pick up at night after the sun sets, it's not nearly as distributive as solar. Large, utility-scale wind makes more sense than microgrids. "On the utility scale wind has beat out solar. For the most part, it's won that competition," explains John Klewin, chief operating officer of Clean Footprint in Cape Canaveral, Florida. "But at the distributive level I don't know wind can hold a candle to solar, which is excellent for rooftops and small ground areas."[30]

Greensburg Mayor Matt Christenson agrees from his basement courthouse office on the windswept plains of southwestern Kansas. Recounting Greensburg's recovery after its devastating

E5 tornado, one we'll hear more about in the next chapter, he notes that smaller wind turbines within town became uneconomical because of their continued service needs. "As mechanical devices, with moving parts that need maintenance, they weren't economically viable to keep running because they didn't produce enough power to offset the cost of that maintenance," said Christenson. "One of things we learned is wind power does not scale down very well. It works great if you want to build a multigigawatt wind farm. But if it's just one or two smaller ones, wind is not going to pay for itself."[31]

We have come far with wind. Onshore wind is competitive with renewables and nonrenewables alike, while future growth will also include offshore wind, which is stronger and more constant. Experts forecast continued efficiency improvements for both. Most turbines on the market today have a capacity between 2 and 3 MW. The next generation, to reach faster wind at higher altitudes, will grow still bigger and boast nearly twice that potential with as much as 4.5 MW. This lowers costs, so future wind farms will likely shrink in size, making wind power more attractive to more expensive real estate markets on the West Coast and Great Lakes. Technological innovation will continue to shape the market, decreasing costs as blades, gearboxes, and generators improve. Governmental investments and regulation will also influence markets as seen with renewable portfolio standards, transmission line expansion, and elimination of fossil fuel subsidies discussed in earlier chapters.

Across this land, we now generate 8.4 percent of our country's electricity from wind, according to the U.S. Energy Information Administration. That's far short of Denmark, the world's leader at 47 percent, but Texas offers promise on what may lie ahead—and a valuable lesson on how to get there. Long our top oil-production state, it now leads the country in generating electricity from wind as well. With climate change in mind, as local conditions allow, the rest of the country should follow its lead. Don't be shy about giving those efforts a boost yourself, by the way. Write local officials. Push for renewable portfolio standards. Sing the praises of wind power to friends, family, and local community organizations. Emphasize not only the need for updating our nation's infrastructure but also the financial benefits that accrue to forward-thinking communities. The times, and winds, are changing.

12

BUILDING (AND REBUILDING) GREEN

Strong southerly gusts slow me to a near standstill as I round the street corner. It's Earth Day, and I'm back in Greensburg canvassing the perimeter of this small Kansan town on a borrowed bike, a gearless model with squishy fat tires that keep me upright and better negotiate rocky, rural roads. Not nearly as old as that used by the Wicked Witch of the West, it's still in the same *Wizard of Oz* genre. And even without a wooden basket to contain Toto, the similarities here in rural, windswept Kansas are not lost on me. With winds at 27 mph, my cartoon-like attempts to power forward better approximate a comical stationary spin session. This is what a true prairie wind feels like. It's no surprise that Kansas is the third biggest provider of wind in our nation, with wind energy potential second only to Texas.

Greensburg is not exactly the Emerald City Dorothy sought, but it is a microcosm of the merits and mistakes when investing in green technology. That's notable in perhaps the reddest portion of an historically red state like Kansas. Koch Industries is headquartered in the closest big city, Wichita. Charles Koch, listed annually among the ten wealthiest people in America, funds several right-leaning organizations, including, along with his late brother David, at least $11 million over the last decade to the Cato Institute and its denial of climate change. And here within Kiowa County, 83 percent of presidential voters preferred Donald Trump in both the 2016 and 2020 presidential elections. That's what makes Greensburg an ideal case to study green tech, that, and the fact that 95 percent of it was destroyed by a devastating tornado over a decade ago.

Tornados typically travel along the ground seventy-five yards at a time. The one that hit Greensburg in May 2007 was an incredible 1.7 miles wide at its base. Centered on Main Street, the storm covered nearly the entire two-mile-wide town. It sat there for eight and a half minutes, submitting inhabitants to astounding wind speeds of 205 mph. Only eleven people died, but 95 percent of the buildings were destroyed. From such tragedy, though, came unique opportunity to change the town's trajectory. Before the storm, Greensburg was struggling to survive, losing 2 percent of its population each year. It was, as *NPR* correspondent Frank Morris describes it, death by one thousand cuts.[1]

"Our situation was unique," explains current Mayor Matt Christenson from his office in the county courthouse basement, the only part of the only public building left standing after the tornado. "It's a lot easier to take big, bold steps when starting from scratch and a clean slate."[2] As the courthouse illustrates, with a tile basement floor constructed from crushed glass bottles and thirty-two geothermal wells three hundred feet below them, bold steps were indeed taken. Tens of millions in financial assistance poured in after the storm. The U.S. Department of Agriculture loaned $17.4 million for the Greensburg Wind Farm six miles southwest of town. The state legislature approved a whopping $32 million aid package. The Federal Emergency Management Agency (FEMA) contributed $80 million in subsidies.[3] Thousands of volunteers also poured in over the weeks and months that followed.

In short, Greensburg had help. It also had the advantage of appropriate scale. Rebuilding with green vision fits well in a small town, one approximating a neighborhood focus. Plus, FEMA and its Region VII staff drew from experience while exorcising some demons from their botched efforts less than two years earlier in New Orleans following the wrath of Hurricane Katrina. By law, though, FEMA funding is restricted to the replacement, not upgrade, of previous facilities, while adopting green technology requires more initial expense than traditional construction.

So why did Greensburg go green?

It wasn't easy. Contentious city council meetings were common in the year following the storm, particularly as citizen frustration built over perceived delays in rebuilding. But eight months after the tornado, the city council adopted a resolution mandating any public building carrying a footprint over four thousand square feet meet something called LEED-platinum standards. That example encouraged residents reconstructing their own homes and businesses to seek higher efficiency as well. Leadership in Energy and Environmental Design (LEED) is a voluntary building efficiency rating program run by the U.S. Green Building Council (USGBC), whose mission is to promote sustainability-focused practices in building and construction. The group began in 1993 and now counts more than six thousand members from a mix of government, nonprofits, academia, and real estate developers.

Launched in 2000, its LEED program is the most widely used green building rating system worldwide and currently in its fourth iteration. Certification is based on six categories: sustainable sites, water efficiency, energy and atmosphere, materials and resources, indoor environmental quality, and innovation and design. Four levels exist: certification, silver, gold, and platinum. LEED is attractive to buyers because it lowers operating costs and raises lease rates, while state and city governments often grant tax credits to companies adopting it. *New York Times* columnist Thomas Friedman says that if he could wave a magic wand and impose one regulation, it would be that every first-year drafting, engineering, and architectural student take a LEED course. "It's a perfect example of not government down, but society up," he contends. Touring the USGBC headquarters in Washington, DC, the first ever LEED-platinum building, I better appreciate Friedman's passion. Elevator lobby walls feature five hundred-year-old salvaged gumwood retrieved from the Tennessee River after the logging industry accidently sank it in the 1800s. High tech HVAC and motion control lighting spread throughout, minimizing their usage. Flooring, walls, and furniture all feature recycled materials. Dual flush toilets and waterless urinals limit wastewater. There's even a two-story waterfall to remove air particulates and provide passive cooling.

Alas, none of this comes without expense. Estimates to achieve certification vary but can add as much as 30 percent in the case of platinum. Paperwork alone can cost a couple of thousand dollars. Some even argue LEED is more public relations than actual sustainability. Critics complain it favors cutting-edge but high-priced technology over common sense, particularly when it comes to residential kitchens and bathrooms. They suggest it's a marketing tool for developers, facilitating premium add-ons beyond the reach of regular folks, akin to the funeral industry's "loading the casket."[14] That's clearly counterproductive when seeking the widest possible adoption of better building practices. "LEED is all about points, and there's some cheap points out there," states Scott Bitikofer, an engineer and former college facilities director who oversaw several LEED projects on his campus. "One of my favorites is, if you put a shower in a building, you get two points. So, for staff, if they bike to work, they can take a shower. That's a great idea. I love it." But LEED's system doesn't consider whether another institutional building 512 feet away already has showers. "That kind of prescriptive thinking frustrates me," Bitikofer adds. "LEED has its place, and it has its place where it probably doesn't fit so well."

Greensburg knows that firsthand. The town built six LEED platinum buildings after the tornado, five of them public. Its 5.4.7 Arts Center was the first building in the state to be awarded platinum designation, while the Kiowa County Memorial Hospital became the first LEED platinum critical access facility in the country. Both provide substantial savings in utility costs: some 70 percent in the case of the arts center and 59 percent with the hospital, according to a National Renewable Energy Laboratory (NREL) study. But both also struggled with alternative energy additions, particularly several smaller wind turbines that failed to withstand fierce Kansas winds. That said, LEED design dramatically reduced each building's carbon footprint, and Greensburg continues to benefit from improved efficiency when it comes to monthly utility bills.

Three critically important factors shaped public opinion on the rebuild, providing political support to go green. For one, media coverage, along with three seasons of reality television programming, created both initial and sustained buzz about opportunities. That also attracted considerable outside interest. "We were kind of the guinea pig, a living laboratory for green technology, green products," says former city manager Steve Hewitt, who now works as chief executive officer of the Kansas Turnpike Authority.[5] For another, the green mantra that investments pay off over time taps into good, old-fashioned thrift and common sense. Greater self-sufficiency in terms of reduced utility needs, thanks to greener appliances and more efficient buildings, also mirrored pioneer independence. Greensburg, for example, was the first American city to light all its streets with LED streetlights. That saves about 70 percent in energy and maintenance costs over older sodium vapor lights. It also reduced the town's carbon footprint by roughly forty tons of carbon dioxide a year.[6] Finally, an ethical and religious commitment inspired better stewardship of the earth. "We were trying to be better than before," said Hewitt. "A smaller impact on the environment is not only about being efficient, but also about taking care of what God has given us."

Town leaders believed that approach could help the town, too. "If you try to build back a town that is dying the same as before, it will just finish dying," stated then Greensburg Mayor John Janssen.[7] Before the storm, youth went off to college and rarely returned, staying in Kansas City or Lawrence for greater economic opportunities. An aging population was left behind. Greensburg, like many rural towns across our Great Plains, had been slowly shrinking for decades, beginning to lose population back in 1960. Many feared that the tornado would be a "coup to grace on a terminally ill town."[8] Greensburg needed a new identity. It recognized sustainability, not only in terms of green architecture and technology but also financially, as a promising one.

Initially, this rhetoric remained complicated, as Bob Dixson, mayor during the majority of the rebuild, explains:

> The term "green" was difficult here on the high plains. We thought green was 1968 and powder blue bell bottom pants with tie-dyed shirts and hair down to here, maybe on mind-altering chemicals. Where in fact, our pioneering ancestors were truly the first green, sustainable people. They lived within their means. We were

> taught: if you take care of the land, it takes care of you. And we were also taught leave her better than you found it. In an agrarian economy, farmers and ranchers know you need a long-term plan. They can't just live year to year. They gotta think about their inputs and how to be good stewards of the land, not just environmental stewardship but financial stewardship.[9]

Energy efficiency can be a low-hanging fruit, attracting converts with future cost savings. Some Greensburg residents, for example, spent more on utility bills than on their mortgages before the tornado, so green buildings and green appliances offered a chance to dramatically reduce those utility expenses. Then again, an argument can be made that emphasizing energy efficiency first is like putting the cart before the horse.[10] The real problem is not how we consume energy but the type of energy we consume. And, as many an innovator discovers, you don't necessarily want to be first. "We want to be on the cutting edge but not the bleeding edge," says Bitikofer. "If you go too soon and get something that doesn't work and doesn't last, then the administration says why are we having to fund this again. That makes it harder to sell the next project. We want to make sure technology has evolved to the point where we can trust it."[11]

That said, those who oppose improving efficiency standards typically overstate the costs to updating while underestimating its benefits. Consumers themselves often lack information about energy efficiency differences among products, since most focus on upfront savings, not savings over a lifecycle. And green tech works best when married with common sense. Take hot water as an example. About 15 percent of the average home energy bill goes to heating water. Turning down your hot water thermostat and insulating your tank are simple steps that complement updating an old heater with a more energy-efficient one. There is even something called vampire loads that green technology can address. Phone chargers and cable boxes, computer cords and coffee pots, all use energy even when turned off, consuming up to 10 percent of household power. A smart grid virtually eliminates that. Unplugging devices not in use does as well.

The biggest source of home electrical use, on average 45 percent of monthly utility bills, comes from heating and cooling. Turning down your thermostat during winter months and up during the summer goes a long way. Drawing your blinds and curtains in hot weather helps, too. There's a reason older structures used window awnings and louvers. Designed to allow sunlight in during the winter when the sun is lower, they also create welcome shade in the summer when the sun is more directly overhead. Green tech complements these steps when smart windows let in more daylight and heat when it's cold while keeping it out when it's hot. Even better, plant some shade trees, replace old single-paned windows with double-paned models, and add insulation.

Knowing the power of Kansas winds and the expense of heating a home during winters on the plains, Greensburg's Debbie and Farrell Allison are big proponents of insulation, namely insulated concrete form (ICF) walls. After losing their hundred-year-old Victorian house in the 2007 tornado, salvaging only a rolltop desk and bookshelf, they built their new home with walls featuring these interlocking, Lego-like blocks of polystyrene foam with six-inch-thick poured concrete in between. Farrell Allison estimates spending $22,000 to $24,000 more for their ICF walls and four cutting edge geothermal wells. That investment began paying dividends immediately, though, saving a couple of hundred dollars each month in utility bills. The couple now essentially makes money from that investment. "People think being green means you're a hippie tree hugger. That's not what it's about," said Allison. "We've tried to make this house the most energy-efficient to reduce our utility costs."

All this requires thinking of buildings as living, breathing organisms. More than a collection of walls, windows, and floors, buildings are a system with multiple interactions.[12] The stakes are high as buildings use roughly 40 percent of the energy Americans consume—and 76 percent of our total electricity. Yet, what incentives exist for real estate developers to choose adequate insulation? Homeowners rarely factor that into their purchase. The same can be said about landlords in the habit of underinvesting in energy-efficient construction, not to mention delaying fixes of drafty windows or installation of more efficient appliances because tenants pay the electric bills. Green tech and the carbon emission savings that come with it are better served if incentive structures reflect utility and emissions costs.

As discussed in chapter 9, utility companies can facilitate investment in solar if they change their business model. Instead of being paid for how much energy they sell to consumers, why not reward utility companies based on how much energy they help their customers save? Discounts could be offered for installing energy efficient appliances or weathering homes to lower consumption. A similar approach is possible with real estate developers and landlords when it comes to building efficiency. Perhaps this means loosening building codes to promote more efficiency or alternative energy investment, such as not counting solar panels or turbines on a roof when reckoning compliance with height restrictions. It may also mean more accurate pricing of day-to-day living costs. The city of Orlando, for example, now requires commercial buildings provide energy benchmarking and reporting. That allows renters to make more informed financial decisions by factoring those utility costs into their renting decisions.

But as Greensburg's experience shows, this works best when pressure comes from the ground up. A financially motivated and committed constituency, as emphasized in the previous chapter on wind, makes a big difference. "You will get a lot farther working with your community and listening to them than basically proscribing rules," notes Greensburg Mayor Christenson. "Without community engagement and buy-in, it's not going to work."[13]

Buildings are not the only energy sector that can benefit from green technology, by the way. Transportation is another, particularly in the United States, where it generates our largest share of greenhouse gas emissions at 29 percent in 2019, according to the EPA. As we shall see, viable public transit is the biggest piece to greener transportation. But cars themselves, like the petroleum that powers most of them, will not likely disappear. Both will become scarcer and more expensive, however. So, the emergence of hybrid, and now entirely electric automobiles, is a critical complementary component.

The world's first commercial hybrid gas-electric car was the Toyota Prius, debuting at the 1995 Tokyo Motor Show. Five years later, the Prius, along with the Honda Insight, hit U.S. markets. By 2012, Prius was among the top fourteen cars sold here in the states. Electric cars, from the popular Tesla series to the Chevy Bolt and Nissan Leaf, increasingly draw attention, pushing the green auto envelope still further. But electric cars occupy only 2 percent of the American new car market today and 1 percent of those on the road. Even as this share will grow dramatically in the next several decades, to 58 percent of new sales globally by 2040, according to one Bloomberg New Energy Finance report,[14] the ever-present danger is that tax dollars will disproportionately support their infrastructure at the expense of public transit. And if EVs are charged by coal and natural gas power plants rather than non–fossil fuel sources, those emissions will be unsustainable when it comes to climate change.

On top of all that, although less efficient trucks and SUVs accounted for only 48 percent of sales in 2012, according to Autodata Corp, that is no longer the case. The calendar year 2019 saw SUVs and trucks comprise almost 72 percent of U.S. new vehicle sales and nearly 76 percent in 2020.[15] Mirroring those statistics, gas mileage for new vehicles dropped in model year 2019 for the first time in five years, causing greenhouse gas emissions to rise by 356 grams per mile per vehicle. Unless rising gas prices reverse customer preferences for SUVs and trucks yet again, electric and hybrid vehicles may be our best hope to arrest this trend.

Both federal and state regulation can be instrumental here. "Government, as a major consumer of services and energy can a be a model, can provide leadership by converting government fleets to electric or hybrid vehicles," notes Paul Owens, president of 1,000 Friends of Florida, a nonprofit that advocates for smart growth in the Sunshine State.[16] California, for example, plans to eliminate new internal combustion vehicle sales by 2035.[17] And the Biden administration recognizes the federal government can provide additional muscle, promising $174 million over the next ten years to beef up our electric car market.[18]

But local demand must drive this. While sticker price, as much as $10,000 more than similar internal combustion alternatives, was long a handicap, range anxiety is now the primary obstacle.[19] President Biden's proposal to build half a million chargers throughout the country by 2030 helps address that.[20] Some simple arithmetic by individual drivers would as well. "This idea of range anxiety is a real concern," says Paul Brooker, senior principal scientist with the Orlando Utilities Commission. "But it's more perception than reality. The typical person doesn't know the average miles they drive in a day. In reality, you don't need three hundred miles. National surveys show 80 percent of people drive fifty miles or less in a day. An eighty-mile range meets 90 percent of people's daily driving needs without recharging. And if you throw in the possibility of recharging at work, we don't even need three hundred miles."[21]

That's the game-changer, integrating electric cars and the power grid. "We have compartmentalized our energy within the transportation community and basically everything else," Brooker continues. "As we electrify our transportation, we are going to see more overlap between those two. And that, to me, represents an opportunity for the grid and building operators as well as vehicle owners. That is an area we are just starting to explore: vehicle to grid, vehicle to home."[22] The future will be built on green technology like this—from cars to buildings to household appliances.

Our Energy Star program addresses this appliance realm with its promotion of energy efficient products. Operating under authority of the Clean Air Act since 1992, it's run by the Environmental Protection Agency and the Department of Energy. The program started with computers and printers, expanding to heating and cooling systems in 1995, and incorporates a range of appliances today, from hot water heaters and dehumidifiers, to microwaves and freezers, to washers and dryers, to dishwashers and refrigerators. In that time, it's saved five trillion kilowatt-hours of electricity, lowering energy costs by $450 billion and reducing greenhouse gas emissions by four billion metric tons.[23] Yet another simple example of green tech at work was phasing out incandescent lightbulbs, 90 percent of whose energy is wasted as heat. Alternatives are admittedly pricier in terms of initial purchase, but much cheaper over their lifetime. Compact fluorescents (CFLs), for example, use 75 percent less energy than incandescent bulbs and last longer. They not only reduce the energy required to produce light; they also reduce the energy needed to cool rooms previously warmed by excess heat from those inefficient incandescent bulbs.

That said, light bulbs and electrical appliances as well as hybrid and electric cars are only a start. Buildings, with their considerable carbon footprint, are where the biggest paybacks lie. "It's the architects who hold the key to turning down the global thermostat," asserts Edward Mazria, a climate-conscious architect from New Mexico.[24] In 2006, his nonprofit, Architecture 2030, issued "The 2030 Challenge" for all new buildings and major renovations to be carbon neutral by 2030.[25] That's the thinking behind NeoCity Academy, a five-hundred-student public choice STEM high school. It is Florida's first high performance school, notes Philip Donovan, studio principal with Little Diversified Architectural Consulting, one hot and humid summer afternoon. Designed to use 76 percent less energy than a typical public school, it's already saving $115,000 annually in energy costs. Total life cycle savings over twenty years should be $3.2 million. Even more important, Donovan hopes NeoCity Academy will be a model for more to come. Like Mazria in New Mexico, he understands the power of local examples. "You champion changing mindsets. Get rid of the phrase, 'This is how we've always done it,' " says Donovan. "It's a grassroots movement that really starts to change the landscape."[26] When it comes to building green, then, demand can better shape supply. It's on you and me to make that happen.

13

ADDITIONAL ALTERNATIVE ENERGIES

We enter the gothic Alaskan ice palace, shivering with anticipation, in the middle of a summer heat wave. Amid majestic ice-carved statues, with a centerpiece ice bar doling out tart appletinis, spread a series of ice-carved rooms. They are mostly bedrooms, but no visitor has lasted a freezing night's stay. From floor to ceiling, the dazzling palace is shrouded in a pallet of pink and violet along with striking slivers of yellow, green, and blue, the colors a nod to the aurora borealis that famously peaks here each March. Chena Hot Springs' Aurora Ice Museum, initially an ice hotel that couldn't keep its clientele through the night, is a cool 25 degrees Fahrenheit, even on a warm July afternoon like today. I've splurged and added an appletini to my admission ticket, saddling up to the ice bar on my ice stool with my well-crafted ice martini glass. Yes, everything in the palace, aside from the lights and liquor, is ice.

That's an impressive feat, especially with this record-breaking summer heat. The museum, along with the rest of this eclectic hot springs resort, much of it salvaged wood, wire, and pipes from Valdez and Prudhoe Bay, is a remarkable testament to human ingenuity and perseverance. The entirety of Bernie Karl's resort has acquired its share of recognition over the years, particularly after being named one of the top one hundred national R&D projects by the U.S. Department of Energy in 2007. This success did not come easily, though. It germinated from failure. Not surprisingly, creativity, risk, and lady luck also featured heavily in the mix. Karl purchased Chena Hot Springs Resort from the State of Alaska in February 1998, a unique opportunity he pounced upon, fittingly, on a Leap Day. He tapped into its geothermal energy to heat his buildings within a week, but the white-bearded entrepreneur of today was no stranger to fossil fuels then. He met his wife years earlier while working on the trans-Alaska pipeline – and his original ice palace was cooled by Caterpillar-built diesel generators, rumbling daily to a tune of $700 every twenty-four hours.

Karl hired the husband-wife team of thirteen-time world champion ice carver Steve Brice and four-time champ Heather Brown to construct that initial six-room Aurora Ice Hotel out of fifteen thousand tons of ice and snow. Within a few months of its December 2003 opening, however, the entire structure melted, and *Forbes* magazine dubbed it the "dumbest business idea of the year."[1] That early assessment may have been on point, but Karl didn't throw in the towel. He learned from his mistake and tried again. "People tell me I'm thinking outside of the box," as Karl recalls. "But hell, I've never been in the box. And I'm never going to get in it."[2] He enlisted help and, after landing a $1.4 million grant from the U.S. Department of Energy, began drilling holes, some thousands of feet deep, to better understand the underlying geology. Within two years, in the summer of 2006, UTC Chena Power Plant fired up as the first geothermal plant in the state. It remains the world's lowest temperature geothermal resource to produce power, servicing the resort's entire electricity needs while saving more than $600,000 a year in diesel fuel costs. Plus, unlike intermittent solar and wind, Karl's geothermal flows twenty-four hours a day.

Geothermal resources heat all the buildings, power the popular indoor and outdoor swimming pools, and even run a hydroponic greenhouse. Ducking into that seven-thousand-square-foot structure, our guide gives us a quick tour. Throughout the year, it produces lettuce, tomatoes, and a handful of fruit varieties for the resort's restaurant as well as its 117 employees. The greenhouse even periodically serves as a testing ground for University of Alaska agricultural products. Geothermal power like this derives from underground reservoirs of steaming hot water generated by friction from continental plates rubbing against one another as well as decay from small amounts of naturally occurring radioactive elements within the earth. First harnessed as electricity to power five Italian lightbulbs in 1904, heat within the upper six miles of the earth's crust holds fifty thousand times as much energy as all the world's oil and gas reserves combined. Geothermal aquifers provide heat to more than seventy countries, although most limit use to bathhouse waters. We lead the world with 3.7 gigawatts installed capacity, enough to power more than one million homes, primarily in central California and western Nevada. That said, we've only tapped six to seven percent of the global potential for geothermal. Here in the United States, the Department of Energy believes that electricity generation "could increase twenty-six-fold by 2050, providing 8.5 percent of the United States' electricity, as well as direct heat."[3]

Geothermal energy is not without risks. You must tap it slowly or risk depletion. Its fluids contain unwanted gases such as methane, nitrogen, carbon dioxide, and hydrogen sulfide, as well as smaller proportions of ammonia, radon, and boron. There are also trace amounts of mercury and arsenic. Geothermal power can endanger thermophile biodiversity, that is, the number and variety of heat-loving plant and animal species. Draining hydrothermal pools can cause soil subsidence, noise pollution, and nasty smells. It can negatively impact pristine views. On rare occasions, deep drilling and fracking even triggers earthquakes.

The main reason geothermal is not more prevalent, though, rests with its high upfront investment costs in assaying and drilling. Knowing where to drill is challenging and expensive. "Wherever we are on the surface of the planet, and certainly the continental United States, if we drill deep enough, we can get to high enough temperatures that would work," explains Jefferson Tester, a professor of sustainable energy systems at Cornell University and a leading expert on geothermal energy. "It's not a question of whether it's there—it is and it's significant. It's a question of getting it out of the ground economically."[14] Ironically, the oil industry already offers an assist here. Some 500,000 oil wells exist in the United States. Alaska is home to tens of thousands of them. Water at temperatures between 248 and 302 degrees Fahrenheit shoot up as that oil is drilled, but this water is usually separated and dumped, costing roughly $4 per petroleum barrel. Imagine if that water was not dumped? What if it was converted into power, as Karl does at Chena Hot Springs?

Here in the Alaskan interior, thirty-three miles off the grid, at both the proverbial and literal end of the road almost sixty miles northeast of Fairbanks, Chena Hot Springs showcases a shift from brown to green energy we would all be wise to emulate. Discovered in 1905 by a rheumatic prospector, this quaint community sits on the "auroral oval" latitude of 65 degrees North, the point at which the earth's upper atmosphere produces the most spectacular northern lights. Asian honeymooners are particularly fond of the resort, which is not bashful about repeating mythology that marriages consummated under the northern lights, as well as babies conceived, will be especially blessed.

Tamping down a different sort of steaminess, the ice museum absorption chiller operates like conventional air conditioners, using the cooling effect of evaporation, along with power produced by the hot springs. Chena Hot Springs waters, for the moment, reach only 165° Fahrenheit, with a nearby creek running a cool 104. As scalding as that still sounds, which we continually remind ourselves of while gingerly exploring the perimeter of a pond and stream on the two-thousand-acre property, it's not ideal for geothermal production. For comparison's sake, consider Iceland, one of the world's major geothermal energy centers, where geothermal water temperatures typically register around 400° Fahrenheit.

While Bernie is drilling deeper in search of hotter water, his current setup still works well. For the ice museum, Karl's hot water heats a mixture of water and ammonia. That mixture, which boils at just 40° Fahrenheit, shifts into pressurized vapor, separates from the water, and enters a condenser where cold water returns it to its liquid state. This water is still pressurized from the energy received from below ground, though, and this high-pressure liquid expands while sucking up heat. Within the Aurora Museum, this liquid brine cools air circulating between an inner and outer wall. All told, the chiller creates fifteen tons of refrigeration per day.

Not every locale boasts sufficient geothermal resources to copy Karl. His example to utilize the local environment creatively, though, is reproducible. The same can be said about his infectious enthusiasm and optimism. "It's pretty amazing what you can do if you think you can," he chirps over the phone.[5] We need more of this mindset in shifting to alternative energies. Alaska is an ideal window into that requirement. Our forty-ninth state has the nation's highest energy costs. Most of its roughly 240 remote Alaskan villages still use diesel generators to supply their electricity needs. Here in the last frontier, winds are ferocious, but melting permafrost compromises the stability of any attempts to construct windmills. Solar power is not exactly attractive in Alaska, either, since power is needed most during the winter when the sun scarcely rises. But Karl's efforts over two decades prove geothermal is a green alternative that can work, particularly in places like Alaska, which holds more geothermal resources than any other state in our union.

Fourteen miles from the city of Unalaska, Karl is working with a group of geologists and geophysicists to bring a long awaited second state geothermal plant online.[6] Geothermal resources were first found on this Aleutian Island roughly forty years ago in the early 1980s. Four unsuccessful wells were drilled before a fifth well hit the jackpot on the eastern flank of the Makushin Volcano. It's what's known as a world class well, with a yield of 15 to 25 MW, much more than what is needed to power all 963 homes in the nearby city. In fact, Karl predicts that strike will anoint Unalaska as the first community anywhere to have zero emissions. "It will be a model for what the world could do," he declares confidently.[7] If successful, the plant will replace 3.5 million gallons of diesel fuel used annually by the main population center of the Aleutian Islands, translating into 39,000 fewer tons of carbon dioxide released into the atmosphere each year.

But again, not every municipality is sitting on a geothermal goldmine.[8] That means considering local context while pursuing alternative energies. And for most Americans, there is nothing partisan about developing such alternatives. "Americans all across the board are wildly enthusiastic about the prospect of clean energy," as George Mason University's Ed Maibach asserts. "The only difference between conservatives and liberals is how quickly this happens."[9] Jonathan Symons, an international relations scholar at Macquarie University in Sydney, Australia, suggests that separating climate mitigation from green values reduces any remaining political polarization regarding alternatives, shifting much needed focus toward development and deployment of zero-carbon technologies.[10] On the other hand, environmentalists like Bill McKibben worry that divorcing green values from energy sources does not solve our climate problem and ignores the potential for more democratic control when economic activity is localized. Culture is central to our climate crisis, his thinking goes, and an emphasis on civil society critical.

Our earlier travels along the home front attest to this, but as Symons suggests, zero-carbon technologies need not be entrenched within other components of climate mitigation efforts. They undeniably overlap, mutually reinforcing one another for many. For others, though, these two objectives appear ideologically incompatible. To link zero-carbon technology to mitigation opens a Pandora's box filled with a supposed liberal regulatory agenda. To borrow from literary giant F. Scott Fitzgerald, recognizing this paradox of diametric takes on zero carbon technology requires holding those two opposed ideas in our minds at the same time. We can pursue alternative energies without any acknowledgement of climate change. And we can pursue alternative energies precisely because there is a climate crisis. Either way, the "test of a first-rate intelligence," to borrow further from Fitzgerald, is fully embracing alternative, renewable energy.

A dichotomous division between fossil fuels and renewable energy is well accepted, but not all renewable energy is carbon-free – and a precise definition of what exactly is renewable remains challenging. Environmental impacts limit universal acceptance of large-scale hydropower, for example, with nearly a third of nation-states around the globe holding renewable energy laws that do not categorize it as renewable.[11] Hydropower generates electricity as water passes simply from higher to lower elevations. But concerns incorporate a complex range of negative impacts, including biodiversity loss, change in local groundwater flows, carbon dioxide emissions tied to decaying plant life behind dams, worsening water quality, enhanced risk of natural disasters in surrounding areas, and, at times, disastrous social impacts from displacing large numbers of people to build the dam in the first place.[12] Dams also collect sediment, which decreases their effectiveness in producing power over time, not to mention reduces the amount of beneficial silt carried downstream to fertilize soil.[13]

That said, most countries concur that energy is renewable when it utilizes natural resources that only deplete at a rate equal or slower than the replacement rate.[14] Despite its negatives, that means hydroelectric power is most often categorized as a renewable alternative. It's an undeniably powerful energy source—and relatively mature technology that produces reliable energy. Across the globe, hydropower is the most extensively used alterative to fossil fuels, registering eighteen percent of the world's electricity. But large hydropower is highly centralized, analogous to utility-scale solar and its democratic hurdles discussed a few chapters ago. Large hydropower also has little potential for growth in the United States, meaning our own search for non–fossil fuel alternatives must continue.

One suggestion targets motion in the ocean. Early technologies date back more than two centuries, with modern designs emerging in the 1960s. The two options here are wave and tidal. Both are vast and dependable as well as largely untapped. Not all waves or tides are created equal, though.[15] Viability depends on geographic location, reinforcing the argument that local conditions should dictate how we diversify our energy portfolio.

More established than wave technology, tidal power exploits high and low tides occurring at regular intervals daily thanks to gravitational pull from the moon and sun. A tidal range of at least 16.4 feet is a prerequisite, and beyond high construction costs, concern about vulnerability to storms slows adoption. The Annapolis Royal Generating Station in Nova Scotia, in fact, is the only tidal generating station within North America, even as we approach four full decades after its construction in 1984.[16] That 20-megawatt power station, though, is one more example of utilizing local characteristics, namely a location in the Bay of Fundy where the difference between high and low tides can be over fifty feet, the highest tidal range in the world. It's not as extreme, but potential exists in northern New England. Ocean Renewable Power Company, based in Portland, Maine, began exploring an option in late 2020, with the goal of developing a primarily tidal generated, $10 million microgrid for the city of Eastport.[17]

Generated by surface winds, wave power is even less explored, but the Electric Power Research Institute estimates it might meet ten percent of total U.S. demand when mature.[18] The best potential, thanks to high waves, lies on western seaboards between 40 and 60 degrees latitude. This includes our densely populated Pacific coast, from Northern California to Alaska. But challenges persist, from design to maintenance to costs. Salt water corrodes equipment. Waves are decidedly more multidimensional than gusts of wind, moving in all directions. As with any construction, damage to marine ecosystems and sea life is a concern. And, along with tidal power, wave power is among the most expensive renewable energy options on our table.

Transitioning back to land, we find a more economical alternative in the form of biomass. With a few caveats, the Intergovernmental Panel on Climate Change (IPCC) defines biomass energy as carbon neutral. Biomass fuels 2 percent of global electricity production, and 20 to 30 percent in states like Sweden, Finland, and Latvia. Alas, most of this comes from wood, as do the majority of 115 biomass electricity generation plants under permitting or construction in the United States. And as Paul Hawken's researchers argue in *Drawdown*, when biomass relies on trees, it is not a true solution.

Biomass energy production essentially taps into carbon already in circulation. As plants grow, they sequester carbon from the atmosphere. Burning biomass releases that carbon, which plants then reabsorb. In theory, with use and replenishment in balance, no new emissions occur. But this depends on using appropriate feedstock. We aren't doing anything close to that right now, relying overwhelmingly on annual grain crops like corn and sorghum. Corn-based ethanol found at your local gas pump commandeers over 90 percent of U.S. biofuel production. Its total carbon footprint, including initial production with fertilizer and equipment operation as well as transit from Midwest production sites to consumers across the country, registers emissions practically on par with those from gasoline. Add to that the fact that corn-based ethanol forces energy and food to compete with one another, and you could easily argue corn-based ethanol is even worse than petroleum.[19]

Native perennial grasses like switch grass or short-rotation woody crops are a different story. They can be harvested several times in a year, for five to ten years before replanting is necessary.[20] Grown on agriculturally degraded lands, native grasses do not displace food production. Instead of damaging biodiversity, they improve it.[21] Less water-intensive, these species also help decrease dependence upon fertilizer-intensive agriculture. That reduces the nitrogen runoff that creates estuary dead zones increasingly seen in U.S. waters like the Gulf of Mexico.

Waste from wood and agricultural processing, similarly, is a welcome feedstock. Scraps from sawmills as well as corncobs, discarded stalks, husks, and leaves from food crops or animal feed all have a lower carbon footprint and avoid competition with food production. One caveat with waste, however, is that developing countries traditionally use different sorts of refuse and biomass for their cooking and heating, including charcoal, animal dung, and firewood. These are deeply problematic for not only the black carbon their burning sends into the atmosphere but also the negative health effects on those who use them. Akin to corn ethanol and its fuel versus food proposition, these forms of biomass often create a fuel versus forest conflict by encouraging land-use change and deforestation.

Complicating matters, our fossil fuel infrastructure continues to restrict growth of alternatives by manipulating markets in its favor. Coal and petroleum have externalized their costs successfully for more than a century. We discussed that with petroleum earlier. Let's take a closer look now at coal. Depending on type of coal, plants emit 1,600 to 2,100 pounds of carbon dioxide for every kilowatt hour of electricity. That's more than double the emissions of a combined-cycle natural gas plant. Burning coal releases sulfur dioxide, which causes acid rain, and oxides of nitrogen, which cause both smog and acid rain. It's also the largest source of mercury, a powerful neurotoxin contaminating fisheries and a danger to pregnant women eating those fish. Its particulates contribute to respiratory disease like cancer as well as cardiac illness. Coal kills thirty-six Americans daily, adding up to more than thirteen thousand a year. Another twenty thousand American heart attacks are tied to coal pollution each year. Annual health costs related to coal exceed $100 billion.[22]

Yet, in 2021, there were still 240 coal-fired power plants across the United States, generating about 22 percent of our electricity.[23] Believe it or not, that's progress, down from 633 coal plants in 2002. Environmental regulations and the emergence of alternatives played a supporting role, but the leading actor here is cheap natural gas, namely the shale gas revolution driven by developments in horizontal drilling and fracking. The market the coal industry long manipulated, ironically, is now escorting it to the dustbins of history. Coal as a source of power has been declining since 2007, and natural gas supplanted it as the primary source of U.S. energy production in 2016, according to the U.S. Energy Information Administration. Digging into the numbers a bit deeper we find that, in 2005, when the technology to extract shale gas became commercially relevant, coal produced half of all electricity generated in the United States. Over the next fifteen years that proportion dropped to a quarter.

This unexpected and often celebrated shift is a double-edged sword, however. The success of natural gas also undermines alternative energy investment, particularly in wind. And the infrastructure built to find and distribute natural gas will not likely disappear quickly, which means natural gas, with its greenhouse gas emissions, will probably be part of our energy mix for decades to come. Given that it emits half the carbon dioxide of coal when burned, natural gas is regularly touted as a bridge fuel. But natural gas is still a fossil fuel, and like its cousins from down below, it's no angel from heaven. A major component of natural gas, and over twenty times more powerful as a greenhouse gas than carbon dioxide, is methane. It's regularly released during drilling. When there is leakage of more than three percent, the climate benefit from burning natural gas instead of coal is lost. This means "gas extracted from shale deposits is not a 'bridge' to a renewable energy future—it's a gangplank to more warming and away from clean energy investments," warns Anthony Ingraffea, a longtime oil and gas engineer who helped develop shale fracking techniques for the Department of Energy, and coauthor of a Cornell study on methane leakage.[24]

A second frequently cited bridge fuel to a zero-carbon future is nuclear power. Nuclear power results from the splitting of atomic nuclei, releasing the energy that holds protons and neutrons together. Like geothermal, this energy heats water to power turbines. Nuclear power generates about 10 percent of world electricity. Its usage in terms of percentage of national mix is highest in France at 76 percent, with the United States relying on nuclear power for 19 percent of our national electrical grid. There are 440 reactors in thirty countries across the globe, according to the International Atomic Energy Agency. Another fifty-four are under construction, with China and India leading the way. Paul Hawken's *Drawdown* team lists nuclear power as their twentieth ranked recommended action in terms of carbon-reducing impacts—although it is the only one of their one hundred suggestions that includes regretful ramifications.

Public concern about safety and environmental impacts from storing nuclear waste hampered expansion of the industry for decades, especially in the United States.[25] It still does. Yet notably, in 2001, the IPCC described nuclear power plants' carbon footprint as "two orders of magnitude lower than fossil-fueled electricity and comparable to most renewables." That's promising and helps explain why a high-profile climate scientist like James Hansen values nuclear in our energy mix. "Renewables like wind and solar and biomass play roles but cannot expand fast enough to deliver cheap and reliable power at the scale the global economy requires," asserts Hansen. "In the real world there is no credible path to climate stabilization that does not include a substantial role for nuclear power."[26] Nuclear production is not carbon-free, though. Fossil fuels are burned in mining, transporting, and enriching uranium, not to mention building the nuclear plant itself, and technically, any technology dependent on an energy source like uranium is not renewable.[27]

Beyond these caveats, three imposing hurdles exist: exorbitant costs, threat of accidents, and disposal of waste. Let's take the last of those first. Nuclear waste from spent fuel remains radioactive for tens of thousands of years. That presents long-term storage challenges and helps explain why, despite approval in 2002 as the national depository for high-level radioactive waste, Yucca Mountain in southern Nevada has yet to open. Advanced generation nuclear power plants, known as Generation 3 and Generation 4 plants, are potential game changers when it comes to this radioactive waste, though.[28] While only five Generation 3 reactors exist worldwide at this writing (none of those yet in the United States), American nuclear proponents already have their eyes on Generation 4 reactors, which are designed to eat their own waste by operating on spent fuel. Second, the threat of accidents, whether due to operator error, equipment malfunction and flawed design, or natural disaster, remains an inescapable handicap. Three of the best-known accidents, the 1979 Three Mile Island partial meltdown in Pennsylvania, 1986 Chernobyl core meltdown in Ukraine, and 2011 Fukushima disaster in Japan, illustrate this all too well. Third, the vast expense of nuclear power alone makes its future uncertain. Energy sources typically fall in price over time. Not nuclear. Nuclear is an astounding four to eight times higher than four decades ago. It costs at least $4 billion to construct plants today and takes, on average, five to ten years, from planning to production, to build them.

It's not surprising, then, to hear the critique of award-winning journalist Naomi Klein, who believes investing in nuclear is nothing more than "doubling down on exactly the kind of reckless, short-term thinking that got us into this mess."[29] Add to this the fact that nuclear power, like large hydropower discussed earlier, is highly centralized and does not tap into the political benefits of locally produced energy. In short, as Oberlin College's David Orr asserts, nuclear power is "slow, expensive, dangerous, incompatible with democracy, and uncompetitive."[30] So where does nuclear power stand over the next decade plus? My best guess is to borrow from the International Energy Agency and its contention that nuclear power will likely play "an important but limited role," provided highly anticipated fourth generation models meet their promises of smaller, lighter, safer, and cheaper production.

Our future energy needs will draw upon the mix of alternatives discussed in this chapter, fluctuating from region to region as local characteristics dictate. One feature will hold constant throughout, though. "The sun is setting on the fossil fuel era," pioneering environmental analyst Lester Brown predicts.[31] That's a bold statement, one that some might find colored by a lifetime's work in defense of the planet. But Brown is not alone. None other than British Petroleum, the oil and gas giant, agrees, announcing in August 2020 that it is abandoning its business model, cutting oil and gas production by 40 percent over the next decade while shifting further into low-carbon alternatives.[32]

Peak oil, passed in 2006 with conventional crude according to the International Energy Agency, and forecast for the 2030s when considering nonconventional extraction, will shape this transition as scarcity works its magic in the marketplace, raising the price of petroleum. Coal, as we have seen with the shale gas revolution, is facing similar market pressures. While the world still holds at least 133 years of proven short ton coal reserves given our current usage rate, coal is losing its long-held, competitive edge. As with petroleum, we will never run out of this resource. We will simply choose more affordable alternatives as it becomes increasingly expensive to use. As Saudi Arabian oil minister Sheikh Ahmed Zaki Yamani famously uttered decades ago, "The stone age didn't end because we ran out of stones."

Alternative energy options are not the only factors shaping our market. Climate threats, increasingly, also adjust our financial calculus. We've seen these mounting costs, from sea level rise to agricultural drought and wildfire threats, increased storm intensity, heat stress, and infrastructure damage. "There's a cost for living on this earth," as Chena Hot Springs Resort owner Bernie Karl warns. "We either pay now or our grandkids pay later, and I believe we should pay as we go."[33] That seems simple. But alas, complications exist. Take the problem of stranded assets. "To slow down global warming, fossil fuel companies must keep 80 percent of their product in the ground, something they surely won't do voluntarily," political commentator Chris Hayes writes. He draws a historical parallel to the abolition movement, "with the same sort of economic impact as having slaveholders free their slaves. That took the worst war in our nation's history," he notes.[34]

Mitigating climate change need not take a similarly violent path—but revolutionary realignment in how we live our daily lives is required. According to the IPCC, low carbon technologies need to generate 80 percent of the world's electricity by 2050 to limit warming to 3.6° Fahrenheit above preindustrial levels. That's a big ask, considering we're at only about one-third today. National efforts like those proposed within the Green New Deal set even more admirable goals of net zero emissions by 2050. But given our current political climate, and traditional American resistance to commands from afar, this tract is more aspirational than realistic. What is already moving our nation forward, though, is on the alternative energy front and bubbling up like geothermal from below, at local levels across this land. We need you to give those efforts an extra political push. As climate scientist Katharine Hayhoe suggests, the best thing you can do is talk about climate change—and talk about solutions like alternative energy.

14

RETHINKING OUR CITIES

My heart skips a beat. Atop an extinct volcano, Mt. Tabor is a lovely city park with breathtaking views of Portland below. The hike up, a popular exercise routine among locals, was a welcome workout. But neither is on my mind now. What has me aflutter is my missing wallet.

I'm here at this Oregonian oasis studying new urbanism, namely how a walkable downtown with mixed housing options, abundant sidewalks, and narrow streets really operates. All this emphasis on building density deemphasizes automobiles and, as I will soon experience, nurtures civic life. It fosters a culture where a stranger from my last bus returns my wallet to the driver, ensuring it reunites with me little more than an hour later. Green architecture, as highlighted in chapter 12, brings much-needed attention to buildings in our fight against climate change. But precisely how these buildings are organized matters, too. This chapter takes this more holistic approach, rethinking design of our cities, including the infrastructure that binds them together.

City design is more important than we often realize. Consider the time it takes to get from home to work. Longer commutes, borrowing from Harvard political scientist Robert Putnam, limit civic engagement and development of social capital.[1] People spread too thinly interact less. Moving them closer together strengthens civil society and decreases political polarization. It also gives us a better chance to address complex societal issues like climate change. Alas, many Americans never consider the social implications of city design. To be fair, most cities, across the globe, were never planned. They evolved in ad hoc patterns. But the older ones, at least their city centers, enjoy the benefit of human-scale design. Returning to this approach will allow cities to reach their potential and lower our carbon footprint. That means rethinking how our cities look. It means planning for people, not cars.

Legendary urbanist, activist, and author Jane Jacobs offers us a recipe. Neighborhoods, and the cities they comprise, work best with a mixture of commerce and residences. Buildings of six stories and less should occupy small blocks with narrow streets but wide sidewalks and plenty of pocket parks. All this facilitates foot traffic and fosters neighborhood interactions that build civil society. It's also good for business. "Sidewalk contacts are the small change from which a city's wealth of public life may grow," as Jacobs poetically wrote.[2]

Walkability, a reflection of density and viable public transit, is not a new idea. Another iconic urbanist, Lewis Mumford, stressed the role of a well-constructed city back in 1938, championing pedestrian-scaled, sustainable cities.[3] Since the mid-twentieth century, though, most American cities have prioritized cars over people. The results have not been good. "Worshipping the twin gods of Smooth Traffic and Ample Parking—[has] turned our downtowns into places that are easy to get to but not worth arriving at," notes city planner and bestselling author Jeff Speck. "Get walkability right and so much of the rest will follow."[4]

Four criteria influence whether people choose to walk, according to Speck.[5] For one, walking must serve a purpose. It must be useful, taking people to nearby workplaces, schools, and shops. Second, walking must be safe. This means wide sidewalks protect pedestrians from car traffic. It also means good lighting at night. Third, walking must be comfortable. Tree-lined and shaded communities like older parts of Winter Park, Florida, near my college campus, do just that. The area also features buildings pushed forward to wide sidewalks with narrow streets and parking in the rear, explains Tim Maslow, former chief planner for Orange County government and current community and economic development director for the city of Groveland, Florida.[6] Finally, walking must be interesting. Describing her Hudson Street home in New York's Greenwich Village, Jacobs labeled this an "intricate sidewalk ballet," one "in which the individual dancers and ensembles all have distinctive parts which miraculously reinforce each other."[7]

More interesting and walkable streets, by the way, are also better for the bottom line. Think about it. Cars don't shop. People do. A study by Transport for London, for example, interviewed shoppers from fifteen London commercial districts. People who drove spent more per individual shopping visit, but those who arrived by foot, bus, or train shopped more often. Collectively, during a month, the last group spent nearly five times more than those who drove.[8] A Portland survey of its commercial districts found similar results. Those who drove to their shopping destinations spent the most per visit, but cyclists spent more overall when counting total visits.[9] Yet our default in facilitating shopping continues to be enhancing access by car. A car-first attitude is more than a drag on profits, though. It hurts our overall health as well. Every minute spent walking extends life expectancy by three minutes, according to one British study.[10] That's a statistic worth repeating. One minute invested in walking returns three more in lifespan.

We used to walk more, at least our children did. Some 50 percent of American children walked to school when I was born back in 1969. Less than 15 percent do now. As we age, with poor walking habits firmly in place, Americans increasingly struggle with obesity, which contributes to coronary disease, hypertension, several cancers, gallstones, and osteoarthritis. Excessive weight kills more Americans today than smoking. As recently as when I graduated from college in 1991, no state had adult obesity rates over 20 percent. In 2007, only one state, Colorado, was under that number. In 2020, not a single state registered adult obesity rates under 23 percent. Compare these statistics to the fact that drivers who switch to public transit drop an average of five pounds. Of course, as Speck cautions, correlation does not prove causation. Suburbs might not make people fat. Heavier people are more likely to prefer driving over walking, so they may be the causal agent, one that shapes higher demand for suburbs.[11]

But cars can strike down our health even more quickly than gaining weight does. From 2008 to 2017, drivers killed nearly fifty thousand pedestrians, according to analysis of National Highway Traffic Safety Administration data by Smart Growth America, a national nonprofit that advocates improving livability of communities by improving transportation options. The most vulnerable are older adults, people of color, and those in low-income communities where wide streets designed for heavy, multilane traffic are more common.[12]

It doesn't have to be this way. Cars don't have to be deadly to those who walk. But "we spent decades building an unsafe system," says Emiko Atherton, director of the National Complete Streets Coalition, a program of Smart Growth America.[13]

An emphasis on driving over walking also contributes to the Balkanization of our society. Suburbanization increases the physical distance between middle and upper middle class versus our poor, making it "much easier and less risky for wealthier taxpayers to ignore problems of those less well off."[14] Those attitudes, with their often-racist undertones, are easily passed on to younger generations. "A child growing up in a homogenous environment is less likely to develop a sense of empathy for people from other walks of life," as architects and city planners Andrés Duany, Elizabeth Plater-Zyberk, and Jeff Speck contend.[15] And putting aside the third rail of race in America, children of all ethnicities lose their autonomy in suburbia. Dependent upon adults to drive them to school, work, or play, they simply cannot practice becoming adults.[16]

Cities like Portland offer an example of what we can do differently. A typical midsize American city, with a population around 660,000, it's the central city in a metro region of twenty-seven municipalities totaling over two million residents spread across three counties. Postwar sprawl brought Portland's metro area the same strip malls and parking lots most of us know all too well. But beginning in the early 1970s, the metropolis started evolving into a model for walking, biking, and public transit. Thanks to a resulting reduction in auto use, Portland is the only U.S. city where carbon emissions today are less than they were in 1990. Central Portland's physical design facilitated this shift with its short, two-hundred-foot blocks. Its economy also weathered transition from shipbuilding, flour milling, and pulp-and-paper processing to producing Intel chips, Nike running shoes, and Columbia activewear, so, crucially, people still live downtown.

Why, you ask? Powell's Bookstore, covering an entire city block, is reason enough for bibliophiles like you and me. Inviting open spaces are another. Jamison Square Park's kid friendly fountain and The Fields Park and its off-leash dog area are just two examples of 275 parks strung across this city. "We need nature where people live, work and play—in the city," explains Mike Houck, emeritus director at the Urban Greenspaces Institute, as we hike through Oaks Bottom Wildlife Refuge, 163 acres of meadows, riparian forest, and wetlands on the east bank of Portland's Willamette River. "The goal is there'll be equitable access to parks, trails, and natural areas throughout Portland and the entire metro region. Greenway and wildlife corridors make that possible."[17]

Even less-green parks can have a positive impact, enhancing social capital, the currency that makes all our lives richer. In the heart of downtown, for example, there's brick-lined Pioneer Courthouse Square, the single most-visited site statewide. Nicknamed Portland's living room, the park replaced a two-story parking garage in the mid-1980s and now serves as a hub for buses and light rail while hosting about three hundred concerts and cultural events annually.

Bruce Stephenson, a professor of environmental studies who has been studying Portland over two decades, is not only an academic proponent of parks like this. He enjoys their benefits every summer and sabbatical from his Pearl District condo. The Pearl District, located just northwest of downtown and sandwiched between Chinatown and Interstate 405, was redeveloped with pedestrians in mind.[18] Three and a half decades ago it was littered with abandoned warehouses along aging factories and known simply as the Northwest Industrial Triangle. Gallery owner Thomas Augustine is credited with renaming it in 1985 after telling a freelance magazine writer the "old, crusty exteriors on the buildings are like the exterior of the oyster shell. But inside it's amazing: There are literally thousands of people inhabiting them, some illegally . . . not only painters and sculptors, but software-makers, wine distributors, poets, and musicians."[19]

Today the Pearl District is home to an enticing array of art galleries and restaurants as well as seven thousand residents and ten thousand jobs. Good bones of short blocks and diverse building structures made this possible. That's what caught Stephenson's eye over two decades ago when he first saw early plans for redevelopment. But the final piece was public transit. "When the streetcar went in, that was the game-changer," he says.[20] Streetcars solve the first mile/last mile problem common to much public transit. They attract more customers than buses, including the occasional user such as visitors like me who often balk at buses because their routes can be confusing. Portland's streetcar service began sharing roads with cars and people in 2001 with a 2.4-mile stretch strategically routed through the heart of the burgeoning Pearl District. Notably, it connected major employment centers like Legacy Good Samaritan Hospital and Portland State University.

Stephenson applauds that decision but is quick to note that continued success is grounded in a grassroots governing system where citizens remain devoted to the public good.[21] Established in 1991, the Pearl Neighborhood Association, for example, is one of ninety-five neighborhood associations in the city. I learn more about its volunteer Planning and Transportation Committee over lunch with its co-chair, Reza Farhoodi.[22] Comprised of fifteen to twenty members who live, work, or own property in the district, it meets the first and third Tuesday each month and is integral to preservation of walkability within the district. That's crucial because neighborhoods like the Pearl District are fundamentally shaped by their walking radius. A five-minute walk, roughly a quarter mile, is a good rule of thumb. "One quarter mile is usually the distance from which you can actually spot your destination . . . and short enough most Americans simply feel dumb driving," elaborate urban planners Duany, Plater-Zyberk, and Speck.[23]

While the Pearl District is a little larger than this, at just under half a square mile, its streetcar service helps stragglers over the hump. Local planning for the streetcar system was integral here, but the state of Oregon gave a big assist decades ago under leadership from Republican Governor Tom McCall. In 1973, the state legislature dictated its metro areas establish urban growth boundaries (UGB) beyond which commercial and residential development were prohibited. This preserved productive farmland and limited suburban sprawl. It's the opposite of drawing lines around a park, as Houck notes. The law also created Metro, the only regionally elected governing board in our country, and it allows cities, every five years, to reassess their land-use needs for twenty years down the road.

Another key measure at the state level occurred a year later in 1974 when Governor McCall told the federal government he would apply funds granted for the Mount Hood Freeway to build a regional public transport system instead. When the Oregon legislature subsequently created the publicly managed TriMet, Oregonians hopped on the transit-oriented development (TOD) train. That was a turning point as TOD promotes density by emphasizing walking distance from residences and business to public transport. "The value in transit is the development you get, not the ridership. The development pattern is what is valuable," asserts Central Florida real estate developer and new urbanism advocate Craig Ustler. "Roads do that, too, by way. They dictate a certain type of development." What makes Portland and its neighborhoods like the Pearl District attractive, Ustler argues, is the "value created not just by assets themselves, but how they are arranged and connected."[24]

That success can come with a price, though, as increased property values often bring gentrification, pushing out lower income families. Homelessness is a related byproduct and a visible problem in Portland, officially reaching 4,015 in 2019 with 8,532 accessing emergency shelters in fiscal year 2017.[25] The city recognizes affordable housing, namely a current deficit of 22,000 housing units for low and moderate-income households, as its Achilles heel, and started to address this deficiency with its Inclusionary Housing Program. It mandates that residential buildings proposing twenty or more units provide a percentage of new units at rents affordable to households at 80 percent of median family income.[26]

Other cities should take note. "Affordable housing is a national crisis," Alberto Vargas, manager of the Orange County Planning Division in Orlando, Florida, warns.[27] Housing is considered affordable when mortgage or rent, along with utilities, total no more than 30 percent of a monthly budget. Yet, in 2019, 37.1 million households (30.2 percent of households nationwide) were "housing cost burdened," spending 30 percent or more of their income on housing, according to Harvard University's *State of the Nation's Housing 2020* report. And 17.6 million in total (one in seven nationwide) were "severely cost burdened," spending half or more of their income on housing.[28]

This brings us back to density. Smart growth is about growing up, not out. But density alone is not enough. To reduce driving and its high carbon footprint, we also need good neighborhood design and viable public transit.[29] Like all politics, this last piece regarding transportation can be intensely personal. "Debates about streets are typically rooted in emotional assumptions about how a change will affect a person's commute, ability to park, belief about what is safe and what isn't, or the bottom line of a local business," Janette Sadik-Kahn, former commissioner of the New York City Department of Transportation, asserts.[30] Cars take center stage here. They are the number one enemy of density, the spreaders of sprawl. And we are the world's most intensive car culture. American households spend 25 percent of our income on transportation. We purchase six million new cars each year. Only 5 percent of Americans use public transportation daily. On top of all that, three-quarters of those who drive do so alone.[31]

Cars were supposed to enhance our independence, to bring mobility and foster freedom of choice. It did for many Americans, for many years, and still does in rural areas. But that's no longer true within most of our cities. Traffic there increasingly restricts mobility. Yet, paradoxically, for many Americans, it's logistically impossible to not have a car.[32] You could say our servant has become our master. It certainly has in terms of time. Our average commute is nearly an hour: 27.6 minutes one-way in 2019, according to the U.S. Census Bureau. Some may enjoy solitude behind their wheel, but time spent commuting means less time for other pursuits and remains a powerful predictor of unhappiness. One German study even found that if one partner commutes longer than forty-five minutes, the couple is 40 percent more likely to divorce.[33]

Then there is the impact on our land. Our nation has 2.6 million miles of paved roads. Add parking to that and you claim more space than the state of Georgia. Every five cars added to the U.S. fleet requires still another acre of land to be paved over, equivalent to a football field, as Earth Policy President Lester Brown details.[34] Parking requirements and their pricing have become the primary shaper of our urban landscape.[35] This isn't cheap. The least expensive urban parking space, an 8.5 × 18-foot rectangle of asphalt on relatively worthless land, costs about $4,000 to create. Those in underground parking garages can reach almost ten times that, up to $38,000 per parking space in San Francisco, says Donald Shoup, professor emeritus of UCLA's Department of Urban Planning and former director of its Institute of Transportation Studies.[36] He calculates the total cost of U.S. parking spaces surpasses the value of our cars themselves.[37]

Driver taxes and tolls pay only about half the costs needed to build and maintain this physical infrastructure, from roads to bridges to parking.[38] Regardless of how little one drives, taxpayers like you and I subsidize the rest. The environmental think tank World Resources Institute estimates U.S. government subsidies of cars, including construction and maintenance of highways, as well as the costs associated with patrolling them, is $111 billion a year more than the taxes paid on motor fuel, vehicle purchases, and license plates.[39]

This die was cast decades ago with the 1956 Federal Aid Highway and Revenue Act. Officially known as the Dwight D. Eisenhower System of Interstate and Defense Highways, and 90 percent funded by the federal government, it launched the largest public works project in world history.[40] Today, that system is a grid of sixty-two limited access roadways that total 47,000 miles. The benefits cannot be denied, but the costs are often ignored. Interstate construction, complete with on-ramps and overpasses, destroyed or divided many inner-city neighborhoods. The system also accelerated our urban demographic migration to the suburbs.[41]

Public transit offers a different path. Spending on it creates twice as many new jobs as spending on highways. Every billion dollars reallocated from roadbuilding to public transit creates 7,000 jobs.[42] Public transit is also considerably less carbon intensive. Greenhouse gas emissions per passenger mile on highspeed trains are roughly one third those of cars.[43] And like in 1956, a case can even be made that public transit addresses critical national security concerns.

But two big problems hold U.S. public transit back. Too often, it's slower and poorer. That explains why, as household income increases, people are less likely to use it. For public transit to take better root we must reverse this relationship. "An advanced city is not one where poor people drive cars, but where rich people take public transportation," as Enrique Peñalosa, the mayor of Bogotá, Colombia, states.[44] Speed of travel will help facilitate that. For now, car travel takes about two minutes per mile for commutes under five miles. In contrast, bus commuting takes more than three minutes per mile—and the average bus commuter waits nineteen minutes just for the bus to arrive.[45]

That's not the case in Portland, where TriMet guarantees minimum frequencies of ten minutes on key bus routes. Older American cities in the Northeast such as Boston and New York also enjoy frequent routes, thanks to a different strategy, subways. By virtue of going underground and not competing with ground traffic, subways are a time-tested way to solve the slow problem. They are, hands down, the most reliable and efficient mode of transit a city can build. But they are not cheap, and something called bus rapid transit (BRT), with its dedicated lanes and express stations, is a worthy and less expensive alternative.

Light rail, like that found in Portland, is another option. Nearly sixty miles of MAX light-rail serve as the backbone for its transit system. Light rail is more expensive to build than buses but costs the same to maintain. And it has something else buses don't have: a mix of nostalgia and newness that attracts users of many backgrounds. Employing low-floored, two-car MAX trains with boxy, angular slanted profiles, Portland's system runs on rails with power from overhead electric wires. These streetcars average 20 mph, including stops, but can reach 55 mph for brief stretches such as the one within a couple of hundred feet of the checked baggage carousels at Portland's international airport. Transit ridership in Portland rivals per capita numbers in Chicago, Philadelphia, and many other older, denser cities.

Portlanders are twice as likely to use public transport as an average American.[46] They also travel 20 percent fewer miles per day, spending far less of their household income on transportation.

Part of why Portlanders spend less is that they bike more. Portland holds the highest bike mode share among sizable American cities, with 5.9 percent commuting trips. This biking, often dismissed as an elite fad in the United States, given our 1 percent national commuting rate, is the final piece to our walkability and public transit puzzle. If every American biked an hour per day instead of driving, the United States would cut its gasoline consumption 38 percent and greenhouse gas emissions 12 percent.[47] Hills and heat as well as cold, snow, and rain present challenges. But more than topography shapes usage, given that hilly San Francisco boasts three times the ridership of relatively flat Denver.[48] Climate does not play the role we might expect, either. Rainy Portland regularly ranks as the second most bikeable city in the United States, with chilly Minneapolis holding the title as "America's #1 Bike City," according to *Bicycling* magazine.[49]

One key to realizing this potential is creating a bike-friendly transport system. Parking-protected bike lanes enhance safety. Beefing up bike share programs like the 12,000 bikes across more than 750 stations within New York's Citi Bike NY enhances access. Giving cyclists an advantage over motorists at traffic lights by allowing them to move out before cars and synchronizing traffic lights to bike rather than car cruising speeds creates incentives. To really move the U.S. needle, though, consider examples such as Denmark, where 18 percent of local trips are on two wheels, and the Netherlands, where it's 27 percent. The world leader in terms of percentage of its people cycling, the Netherlands has more bikes than people, 22.5 million compared to 18 million. In central Amsterdam, nearly half the traffic, at

48 percent, is by bike. Denmark's capital of Copenhagen is no slouch either, recording rates of 36 percent.

Amsterdam and Copenhagen weren't always bike-friendly. Both cities made their streets car-centric after World War II. But the people demanded better, and public protests surrounding traffic deaths instigated shifts in the late 1960s. Amsterdam responded by segregating bike lanes from cars. They disincentivized car ownership to make biking more attractive and made infrastructure for cars less accommodating, gradually removing parking spaces. Of course, there's also a cultural piece here. Europeans typically shop for food daily, for example, instead of weekly as Americans do. Bike trips naturally lend themselves to errands like that. The truth is that real advance happens not just when we bike more individually but when we advocate for high density development and biking infrastructure that allows others to bike more as well. That's the path Portland pushed decades ago, one that continues to pay dividends today.

All this adds up to more biking, public transit, and walkability options at the local level, which requires fundamentally rethinking how our cities operate. John Muir famously eschewed city living as toxic, pushing for national parks as restorative retreats free from urban blight. And until the last century, large American cities regularly struggled with pollution and disease, their inhabitants passing away ten years sooner than rural residents in 1880, for example.[50] Modern medicine, sanitation advances, and environmental regulations, though, reversed that pattern as U.S. life expectancies in urban areas surpassed those in rural communities in 1940.[51]

Cities are much more than a health asset today. Even as technology continues to facilitate telecommuting, cities remain a center of creativity and cultural amenities. And with climate change, cities are key to saving our planet, as former New York

City Mayor Michael Bloomberg asserts. He lists four reasons. For one, cities are on the front lines of climate change, accounting for about 70 percent of carbon emissions. For another, city mayors are less ideological than national legislators. They deliver essential resources and solve local problems, as Bloomberg points out, so they must be pragmatic to politically survive. Third, mayors recognize addressing climate change is a medium to speeding up economic growth. Fourth and finally, as we have seen throughout this chapter, cities offer scale thanks to their density.[52]

Rethinking our cities requires a two-pronged approach.

On the transportation end we need a Copernican-like revolution, as Sadik-Kahn argues. The struggle here is not with the science of climate change. It's not even with twentieth-century traffic engineers championing roads' sole purpose as moving cars through and often out of a city as quickly as possible. "It's within the culture and idea of whom streets serve," as she states.[53] American cities and their streets were made for cars. But that was a different time, and this century requires a different approach, one where people are the focus and walkability with public transit emphasized.

That works only in areas of sufficient density, our second prong. Targeting density means we must better emphasize diversity. So many challenges across this land come from misinformation, from lack of exposure to different people and different ideas. "Does anyone suppose that, in real life, answers to any of the great questions that worry us today are going to come out of homogenous settlements?" urban activist Jane Jacobs once asked.[54] Jacobs was emphasizing infrastructure, but her target was people themselves. She identified four generators of diversity for a neighborhood district, contending that all four were necessary and stressing them as central to her seminal book.[55]

First, neighborhoods need more than one primary function, and preferably more than two. Second, blocks must be short. Third, buildings should range in age and condition. And fourth, there must be density.

Portland neighborhoods like the Pearl District get this. They understand that's what is needed to attract and retain talent. As American urban studies theorist Richard Florida argued two decades ago, twenty-first-century cities will flourish only if they entice such creative people.[56] That also gives us a fighting chance when it comes to climate change. Still, while Portland offers us a playbook on how to better evolve with this in mind, continued progress across this land depends on you—and whether you push for similar changes in your hometown.

15

LIVING WITH CHANGE

"That's it?" my teenage daughter grumbles. "That's all?" Admittedly, the legendary fountain of Ponce de León fame is underwhelming. That's partly because our expectations are fed more by fiction than fact, and partly because nothing could really impress us with this sweltering combination of Florida summer heat and humidity amplifying our orneriness by the minute. We're exploring old Florida this weekend, sampling attractions established long before Disney and its theme park cousins like Universal and SeaWorld made our state an international tourist destination. St. Augustine's Fountain of Youth Archaeological Park fits that bill well, from a carefully crafted Timucuan village recreation to an authentic Spanish colonial-era blacksmith demonstration and cannon firing to the pretty peacocks roaming throughout the fifteen-acre grounds. With a guestbook stretching back to 1868, the park professes, with good reason, to be Florida's oldest continuing tourist attraction. The same veracity cannot be attached to its namesake, namely the legend that those who drink or bathe in its waters restore their youth.[1]

It's an unassuming stream of alkaline water that trickles out of the staged diorama before us. Filtered through layers of limestone and laden with more than thirty harmless minerals, about half of which my mind imagines it can taste, the water is restorative only in the sense that we are hot, sweaty, and thirsty. Probably the main turnoff, at least for my youngest child, is the smell of sulfur. But that hasn't stopped countless others before us from seeking out these waters.

Beware that hype, though. Everything is not always as advertised. Historians believe Ponce de León never walked these grounds, let alone drank from these waters. His landing in St. Augustine, they suggest, was invented to snare tourists like me generations ago. It's an ingenious marketing idea, tapping into our cultural obsession with youth. But resistance to aging, resistance to change in general, can get us into trouble. Relying solely on adaptation to counter climate change follows a similar flawed logic. Yes, we must learn to adapt to change. Yes, adaptation must be part of our climate policy tool bag. But just as important, we must respect limits to those adaptations—and the added risks they may present.

St. Augustine is prime example as to why.

Ponce de León might not have tasted these waters, but he did land along Florida's coast somewhere between the mouth of the St. Johns River and Cape Canaveral in April 1513. That was the Easter season, known as the feast of flowers in Spain (*Pascua Florida*), and thus the name La Florida. The Spanish conquistador and explorer proceeded southward along our coast and discovered something he wasn't looking for, the northerly Gulf Stream that we noted back in chapter 1.

Most historians believe he wasn't searching for the fountain of youth, either, since no record of such a search appears in his writings. The legend first arises in the 1800s when American writer Washington Irving, author of another popular myth mixing fact with fiction, "The Legend of Sleepy Hollow," concocted the fountain tale that history books and tourist attractions would often repeat.[2] Granted, a settlement was founded here by the Spanish, albeit several decades later in 1565 by naval commander Pedro Menéndez de Aviles, making St. Augustine the oldest continuously occupied European settlement in the United States. It's also our nation's first planned town, built according to the Laws of the Indies as laid down by Spain's King Philip II. These 148 ordinances spelled out in detail where and how to build new settlements. To this day, the legacy of those regulations maximizes breezes and shading on St. Augustine's streets and makes the old town's 144 historical blocks a pleasant place to stroll, even as hordes of tourists regularly descend upon them.[3]

This mix of cultural and historical character positions St. Augustine squarely within our adaptation discussion. Sea level rise and increased storm intensity tied to climate change threaten its buildings, streets, and people. Warmer temperatures from climate change also mean more intense rain events, and St. Augustine struggles with flooding as a result.

But remarkably, it doesn't even have to rain for St. Augustine to flood. High tide swamps its low-lying streets twelve to sixteen times a year. "We locals like to joke, 'You can spill a glass of water and it will flood in St. Augustine,'" laughs Athena Masson, a meteorologist for the Florida Public Radio Emergency Network and city native.[4] St. Augustine is precisely why, esteemed Princeton professor Michael Oppenheimer asserts, "Policymakers no longer have the luxury of downgrading adaption, because climate change's devastating effects are no longer in the future; they are occurring now."[5] This city shows us adaptation and mitigation are not an either/or proposition. Powerful synergies exist between the two. Adaptation, when done correctly, complements mitigation while failing to incorporate mitigation dramatically increases adaptation costs.

That last piece is important. Adaptation cannot come at the expense of mitigation. Mitigation reduces exposure. It reduces vulnerability. It makes adaptation more realistic. "Adaptation and mitigation are complementary strategies for reducing and managing risks of climate change," as the Intergovernmental Panel on Climate Change (IPCC) summarized in its 2014 Summary for Policymakers. "Substantial emission reductions over the next few decades can reduce climate risks in the 21st century and beyond, increase prospects for effective adaptation, reduce the costs and challenges of mitigation in the longer term and contribute to climate-resilient pathways for sustainable development."[6] But again, adaptation, like the fountain of youth, is not always what we wish it was. "Adaptation can reduce the risks of climate change impacts, but there are limits to its effectiveness, especially with greater magnitudes and rates of climate change," as the IPCC warns.[7]

Adaptation occurs at many levels, from the international and national to the regional and local. Most important, though, adaptation is place and context specific, with direct impacts a product of local geography. The basic tenet is that actions now should increase options available later. Elected officials typically are more comfortable talking about adaptation because it emphasizes action. Also, by avoiding discussion of underlying climate change causes, adaptation is less politically polarizing. From Florida to Alaska, scientists, activists, and governmental officials echo this sentiment. "My work is in adaptation rather than mitigation. Adaptation strategies tend to be a lot less divisive," explains Nancy Fresco, SNAP network coordinator and associate director of the Cooperative Institute for Alaska Research at the University of Alaska Fairbanks. "The frontier plays very much within the sensibilities of Alaskan culture. Adaptation strategies fit in well across the political spectrum. The thinking is whatever happens we'll tough it out."[8]

Adaptation taps into the American ideal of rugged individualism and resilience.[9] This translates directly into financial savings, with every dollar invested in resilience saving four in avoided costs, according to the National Institute of Building Sciences. Cities cannot do this on their own, though. Resilience requires coordination with multiple local and regional government agencies.

A good example is further south down the Florida coast. The Southeast Florida Regional Climate Change Compact, the partnership between Broward, Miami-Dade, Monroe, and Palm Beach counties noted back in chapter 1, seeks to reduce the regional carbon footprint, implement adaptation strategies, and build climate resilience.[10] That begins with access to updated information such as climate indicators like sea level rise, nuisance flooding, saltwater intrusion, the public health/heat index, sea surface temperature, severe storms, precipitation, drought, and plant hardiness. This data then drives projects such as those in Miami Beach where raised streets and improved storm drainage brought immediate relief. But these are only temporary measures, and we cannot allow initial successes to mislead us into thinking our problem solved. This is a constant risk with adaptation, that it reduces political pressure for continued mitigation. Perhaps most dangerous of all, as environmental activist and writer Kathleen Moore asserts, the risk with adaptation is "we accept living with the destruction we are causing—and the moral failings that characterize climate change itself become replicated and amplified in many of the plans to adapt to it."[11]

St. Augustine is just one example of why we must resist this temptation. Although downgraded from its original Category 5 strength in the Atlantic basin, Hurricane Matthew was still a Cat 3 storm when it brushed offshore in October 2016. Its record-breaking nine-foot storm surge was not the first hurricane to damage the historic city, and it won't be the last.[12] With low-lying critical infrastructure, St. Augustine has always been vulnerable, well before the United States even purchased it from Spain along with the rest of Florida in 1821. From the start of Spanish settlement, in 1565, a hurricane provided strategic cover for Pedro Menéndez de Aviles and his army to attack the French settlement at Fort Caroline, solidifying Spanish control of the region. In 1599, a hurricane sheltering island in the harbor disappeared while a sandbar shifted to make the inlet shallower. In 1605, yet another hurricane flattened and flooded much of the city. And that doesn't even cover half a century.

Seventy years later, a 1675 description by Bishop Gabriel Díaz Vara Calderón of Cuba refers to St. Augustine as "half submerged from hurricanes as it lies at sea level."[13] Indeed, it only takes three feet of storm surge to swamp the historic city today, explains Reuben Franklin, public works director for the City of St. Augustine. "Defending looks pretty costly for a small community our size to tackle three feet," he admits.[14]

It's not that St. Augustine hasn't tried. A series of downtown drainage improvements covers a basin area of more than eighty-six acres. Officials have raised the street profile near the Bridge of Lions, added pumping stations in several flood-prone areas, and retrofitted all storm water drainage with backflow prevention valves. But sea level is still rising, and hurricanes, along with their storm surge, are becoming more intense. In 2016 the University of Florida produced a report examining the impacts from sea level rise in the city, looking at one, three, and five feet over the "mean higher high water" mark, which is an average of high tides over the years. They found that sea level rise would range wildly, between 0.25 to 6.67 feet, over fifteen to eighty-five years.[15]

St. Augustine recognizes its continued vulnerability in this context. Extreme weather events like Hurricane Matthew, centered thirty miles off the coast, focused public attention.[16] The damage is still visible. Magen Wilson, executive director of St. Augustine's Historical Society, explained to me that storm waters were one and a half to two feet high in their complex after Matthew and stayed that way for three days before receding. Coquina buildings, constructed with soft limestone of broken shells often used for roads in Florida and the Caribbean, held up well. Those with drywall did not. While cleanup took weeks, an even bigger problem for the Historical Society was damage to the landscape. "Saltwater comes in and sits here," Wilson explains. "We are still removing trees. Things we thought made it are still dying. Two more trees must be removed later this week."[17]

A few weeks after that conversation, I was invited by St. Augustine's city manager, John Regan, to meet with his public works staff at the annual American Public Works Association meeting. Several St. Augustine employees were presenting talks on storm water management and emergency preparedness, and the opportunity to hear from professionals across the country was a welcome respite from my usual late afternoon grading. The presentations were nothing like the stereotype engineers engender. They were engaging and mostly refrained from jargon incomprehensible to a layperson like me. My biggest take away was that St. Augustine will always require adaptation measures, and, at some point, those measures will challenge the very characteristics that dictated adaptation in the first place. "Our existential conundrum is: Can St. Augustine adapt while continuing to embody those features that are fundamental to its current character?" as then city program manager Reuben Franklin asked.[18]

In St. Augustine, like many other places, this discussion centers upon the conventional approach to seaside adaptation: seawalls. It's not a new idea. Ancient Greeks and Romans used them. The Ottomans did, too. But they don't work everywhere. Farther down the peninsula in South Florida, as addressed in chapter 1, seawalls do nothing to stop salt water from seeping up from below through the porous limestone bedrock found there. Seawalls also can do more damage than they prevent. They harm valuable marine and shoreline vegetation. They restrict natural beach responses to bad weather, preventing transfer of sand between dunes and surf zones. And they give a false sense of security, encouraging construction where it should not occur.

With that in mind, let's take a closer look at St. Augustine's seawall. The Spanish built the city's original seawall between Castillo de San Marcos and the Plaza de la Constitución from 1695 to 1705. City archaeologist Carl Halbirt even found residue from this wooden wall, three inches thick and hammered into the riverbed. By the time Florida became a U.S. territory in 1821, that original wall had largely deteriorated. A new seawall, roughly seven feet in height and made with coquina, replaced it. The U.S. Army Corps of Engineers extended that wall farther south between 1835 and 1842, capping it with granite to protect the coquina rock from the wear of people strolling upon it. "Citizens wrote a letter to the federal government asking to maintain the Castillo in 1834 under the guise of preservation," explains Susan Parker, former executive director of the St. Augustine Historical Society. "But really, their motivation was more about improving the sea wall for protection."[19]

Fast-forward a century, and the seawall faced additional development threats, namely from road widening and parking. Between December 1958 and July 1959, stretches succumbed to a four-lane highway (A1A) skirting the eastern edge of town, despite local and national preservation group protests.[20] Between 2012 and 2014, after a portion collapsed during Tropical Storm Gabrielle in 2001 and considerable flooding from Tropical

Storm Fay in 2008, a new 1,200-foot reinforced concrete seawall was built thirteen feet farther out into the bay. It's a lovely promenade, with views of the historic coquina seawall behind it, but that $6.7 million project continues to extend development into the water instead of away from it—and only protects the city from Cat 1 hurricanes.

City officials understand this predicament. "By the way, seawalls are not the future,"[21] John Regan cautioned over lunch within sight of the Bridge of Lions and Mantanzas River. With over two decades of service to the city, first as utility director and then as chief operations officer before assuming his current position, Regan should know. "We need to reimagine our infrastructure," he asserts. Part of that reimagination involves some difficult and costly political decisions. Truth be told, the most effective adaptation to sea level rise is managed or planned retreat. It's also the least popular. David L. Kelly, professor of economics and director of the sustainable business program at the University of Miami, highlights the main reason why. While most adaptation measures to residential properties, 145 of the 158 he studied, bring small but significant financial benefits to the property owner, retreat does not. Kelly found, at least in his South Florida sample, that moving two-thirds of a mile farther from water decreased property values by 1.1 percent. In comparison, a one-meter elevation in property registered a 2.8 percent increase in value.[22]

This study, though, says nothing about the value of a property that does not move but is flooded. While retreat is by no means cheap, not retreating, in the long run, looks to be even more expensive. "It's just a sad fact that we can't spend an infinite amount of money defending the coast," Oppenheimer cautions. "And the concept of retreat, which is sort of un-American, has to be normalized. It has to become part of the culture. Because there are some places where we're really going to have to retreat."[23] Jim Cason, senior advisor to the American Flood Coalition and former three-time mayor of Coral Gables, agrees and points out we best start setting aside funds to support that retreat now. Even within sections of cultural and historical treasures like St. Augustine, it's inevitable.

That leaves two key questions: Will this retreat be sooner or later? And how will we retreat, proactively with concerted planning—or retroactively after disaster?[24]

While retreat is often interpreted as defeat, the truth is a bit more complicated. "We need to look at where we can come up with natural solutions," asserts Cason. "We need to replant mangroves, put dunes back in."[25] Cason is talking about something called a living shoreline, providing more permeable borders for intertidal habitats like wetlands and salt marshes to survive and flourish.[26] While about 14 percent of U.S. coasts are "armored" with hard infrastructure like concrete seawalls, NOAA holds that this too often destroys habitat and causes shorelines to erode even faster.[27] It makes more sense, particularly in sheltered tidal areas, to stabilize banks with wetland plants, sand dunes, stones, and other natural elements like oyster reefs, coral reefs, and mangroves.

How exactly do you switch from a developed to undeveloped coast, though?

In extreme circumstances, states can condemn private property with the power of eminent domain, but the fairest solution really is to buy them out. New York, for example, spent $240 million to buy 610 properties on Staten Island after Hurricane Sandy hit in 2012.[28] Some 145 families in the Oakwood Beach neighborhood banded together and sold their houses. The state knocked them down and returned the land to nature.[29] But costs here can be considerable, and buyouts really only work when an entire neighborhood agrees.[30]

Yet another controversial strategy when it comes to adaptation revolves around geoengineering. Most ideas here are little more than wishful thinking. They sound attractive, like the fountain of youth, selling an "intoxicating narrative that technology is going to save us," as best-selling author Naomi Klein asserts.[31] But in truth, as Penn State climate scientist Michael Mann and his coauthor, cartoonist Tom Toles, contend, adaptation "merely provides crutch[es] for critics of restraints on carbon emissions," reinforcing the misconception that there is no need to break our carbon addiction.[32] The fundamental flaw of geoengineering adaptation measures, in fact, is twofold. For one, they fail to deal with our core problem, too much carbon dioxide. For another, they are risky experiments conducted on a grand planetary scale, which means they might create as big a problem as they attempt to solve.

Solar Radiation Management (SRM), the spraying of sulfate into the stratosphere to reflect sunlight, is one example. This occurs naturally when volcanoes erupt. Most volcanic eruptions send ash and assorted gases into only the lower atmosphere, but stronger eruptions such as Mount Pinatubo in the Philippines shoot emissions higher, into the stratosphere. When it released twenty million tons of sulfur dioxide in 1991, circulating around the entire planet for roughly eighteen months, those droplets became miniature light-scattering mirrors, reflecting the sun's heat before it reached the earth's surface. Global temperatures dropped nearly a degree Fahrenheit.

The technology to replicate such natural SRM is not the tricky part. It's the expense, namely $2.25 billion per year for the initial fifteen years.[33] Even more troubling, these estimates fail to consider unintended consequences such as changes to Asian and African summer monsoons and their life-giving precipitation that grows food for billions. Neither do such estimates say anything of the permanent haze that would cover our sky, reducing solar power capacity around the globe. These cost estimates also ignore continued losses from ocean acidification, which SRM does nothing to prevent.[34] Similarly, since SRM addresses only the symptoms of climate change and not the overarching problem itself, once chosen as a "solution," we are married to spraying sulfates into the atmosphere—forever. Our species is not so good at the forever thing. On top of all this, SRM raises disturbing political questions about oversight. That was the plot of the 2017 Hollywood film *Geostorm*, and it's not far-fetched to imagine life imitating fiction, at least when it comes to political fights over who controls such a system.

Carbon Capture and Sequestration (CCS) is another questionable adaptation measure. CCS is limited geographically. Storage requires geological formations like impermeable caprock, saline aquifers, or oil-bearing strata that allow natural chemical processes to solidify the carbon dioxide. CCS is also expensive. With 20 to 25 percent of the energy produced needed to power the actual station, its electricity simply cannot compete with alternatives like natural gas, and, increasingly, thankfully, solar and wind. I toured a prototype of one outside Berlin, Germany, in 2009. It's no longer online today precisely because of that competition.

These flawed examples aside, promising adaptation strategies do exist, from our seashore to the heartland. Sometimes this can be as simple as painting our infrastructure white, from roofs to parking lots. Even mountains might benefit from some white paint. The World Bank listed that as one of its hundred ideas to save the planet back in late 2009, a nod to Peruvian inventor Eduardo Gold's attempt to resuscitate the lost glacier at Chalon Sombrero by reflecting heat and cooling the mountain's summit.

Artificial trees are another creative idea. These leaves capture sunlight and convert it to wireless current. Klaus Lackner, director of the Lenfest Center for Sustainable Energy at Columbia University, believes that these artificial "leaves" can be one thousand times more efficient than natural leaves, with ten million trees removing 3.6 billion tons of carbon dioxide per year. That's roughly 10 percent of what we pump out globally each year. "We don't need to expose the leaves to sunlight for photosynthesis like a real tree does," Lackner explains. "So, our leaves can be much more closely spaced and overlapped—even configured in a honeycomb formation to make them more efficient."[35]

Then there is the long-standing natural tree planting option, often referred to as carbon offsetting. These efforts are not always maintained, though, which means that trees planted may not survive and mature to become viable carbon removers. Carbon offsetting is also vulnerable to exaggeration and fraud, as seen in each of the offset mechanisms the 1997 Kyoto treaty created, the Clean Development Mechanism (CDM) and Reducing Emissions for Deforestation and forest Degradation (REDD). The Stockholm Environment Institute, for example, estimates that 75 percent of CDM funding paid for actions that would have happened anyway, and that only 25 percent of money spent showed real dividends in carbon reduction.[36]

Borrowing from the last chapter on how city street design matters, and looking again to St. Augustine, provides additional insight on adaptation. Narrow street width in warm climates provide those on foot with much-needed shade. Curving streets create additional shade opportunities and act as flues funneling air flow down them. St. Augustine is a model to emulate in both respects. But when it comes to flooding thanks to sea level rise and storm surge, city officials face something of a Sisyphean task.

St. Augustine is not alone. All along our coast, we need to better live with water. This means better adapting how we build near water. Replanting trees and rebuilding dunes, which act as natural storm buffers, is a start, particularly as developers often remove them to improve ocean views and raise selling prices.[37] Better yet, rethink construction altogether. Until federal flood insurance became available in 1968, beachfront buildings were difficult to insure against flood damage. That meant that beachfront property owners had to pay cash for their property—and that most structures were much more modest than those today. People understood that investing more was too risky.

Adaptation in the face of climate change, of course, is not always about too much water. Sometimes the problem is not enough of it. As we saw in chapter 3, considering increasing threats from wildfire, creating more defensible space in our wildland urban interface is another key adaptation strategy, albeit one that should be used in tandem with retreat. Furthermore, in terms of agricultural adaptations in drier climates, those can be as simple as converting irrigated agricultural land to solar and wind use. Solar panels, in fact, make good economic sense on at least 9 percent of idled land, according to the California's Public Policy Institute.[38]

Another adaptation involving agriculture targets livestock. The EPA estimates that agriculture contributes 9 percent of U.S. greenhouse gas emissions. Cows, sheep, and goats produce more emissions relative to pigs and chickens. Changing this balance could easily be labeled a mitigation measure, but it obviously involves adaptation for American palates and plates, too. No-till farming is still another agricultural adaptation/mitigation worth pursuing. Soil is our second-largest carbon repository, storing nearly twice as much carbon as plants and the atmosphere combined. As the Environmental Defense Fund's Miriam Horn suggests, no-till farming "helps farmers adapt to climate change by putting back into soils carbon lost to the atmosphere through cultivation."[39]

Here in Florida, with 1,350 miles of coastline, the adaptation stakes are high. "We know we are in for two or three feet of sea level rise no matter what we do," warns Susan Glickman, Florida Director of the Southern Alliance for Clean Energy. "Even if you turn the carbon dioxide spigot off tomorrow, you can't adapt your way out of this problem."[40] Adaptation is not our fountain of youth, not our silver bullet to solve climate change. But adaptation does buy us time to tackle the root causes of our climate crisis. And the more mitigation we do, the less adaptation will be required. "We basically have three choices: mitigation, adaption, and suffering," asserts John Holdren, science advisor to former President Barack Obama. "We're going to do some of each. The question is what the mix is going to be."[41]

CONCLUSION

Think Local, Act Local

Climate change is a complex problem, with lots of moving parts. Yet its solution is disarmingly simple: we must change how we interact with the planet. Awareness of environmental interdependence is critical here, for human and environmental health are deeply intertwined. Alas, the degree to which this message is heard depends largely upon the messenger. Best-selling author and journalist Malcolm Gladwell highlighted this in his work on tipping points.[1] It matters where we get our information. As scholars Riley Dunlap, Aaron M. McCright, and Jerrod H. Yarosh demonstrate, in survey after survey for over two decades now, that's particularly true with climate change.[2]

Why not cut out the middleman, then? Make yourself the messenger. Step outside your comfort zone and see this land yourself. This book is meant to inspire such local adventures, expressly targeting those that recognize climate change as a problem but struggle finding motivation to act. As the Yale Program on Climate Change Communication suggests, Americans tend to view climate change as a distant problem. Surveys show 61 percent of Americans say climate change poses a risk for us, but only 43 percent think it will affect them personally.[3]

We consider it temporally, spatially, and socially removed from the here and now.[4] We see its threats as impacting future generations alone, not to mention places far away.

The first half of this book demonstrates otherwise.

Edward Maibach, university professor and director of George Mason University's Center for Climate Change Communication (4C), draws on more than a dozen years of climate communication studies when he emphasizes this point. "We want people to understand climate change is a here, now, wherever we live in America problem," Maibach explains. "Past public policy successes feature a simple message, repeated often, through a variety of trusted voices. At my center, we try to build on this formula. We are focused on reducing that psychological distance, helping Americans understand this is our problem now."[5]

Not surprisingly, psychological distances dissolve when we establish more personal connections, when we see climate change impacts on this land we call home. There's intuitive logic to this. People, put simply, are nicer to their neighbors, to people they know. People also show more respect to the inanimate objects they know, like the forests and seashores they visit. That's why support for national parks is high regardless of a traveler's political persuasion. And it's why I trumpet the power of direct experience, including eco-conscious travel. Again, this travel need not be far. While all travel comes with greenhouse gas emissions, carbon costs shrink as green techniques are applied. It matters how far you travel and what mode of transportation you use. But it also matters how you behave at your destination, and, perhaps most important, how you adjust daily life afterward.

Travel is not all good—but neither is it all bad. True ecotourism, on the other hand, can be an invaluable asset in our emerging climate action tool bag. Not only does it, by definition, minimize one's carbon footprint. It enhances local decision-making power by putting money directly in the hands of those who live where we visit. Those funds then better balance economic and environmental interests, adding tangible economic value to ecosystems locals previously exploited with only short-term economic interests in mind. Finally, ecotourism, and by extension eco-conscious travel, holds the potential to change minds about our climate crisis. Experiences closer to home resonate most effectively here. "There is a lot of evidence behind the idea that personalizing climate change and helping people understand the local impacts are more important than talking about how it's influencing melting glaciers or talking about wildfires when you live in Ohio," as Jennifer Marlon, a research scientist and lecturer at Yale, states.[6]

Conscientious travel offers a bridge over another divide as well, one that threatens the foundation of our democracy. Increasing political polarization runs much deeper than what former President Trump exposed and encouraged. His administration accelerated erosion of our democratic norms, but it didn't give birth to it. Harvard professors of government Steven Levitsky and Daniel Ziblatt illustrate a decades-long decline of the "soft guardrails" of American democracy, with modern roots in extreme partisan polarization during the 1980s and 1990s.[7] Polarization of elected officials began capturing academic interest over two decades ago, with scholars suggesting that Congress is more polarized now than any other period in the last 125 years. Levitsky and Ziblatt focus on elites, noting that the mass public is typically less polarized.[8] As such, top-down strategies targeting Congress or political parties, especially Republicans and their rightward drift, may hold a bigger bang for our bucks.

But focusing on changing institutions and their rules, a combination most scholars often target, will take time. That's time that we don't have, as the former director of the NASA Goddard Institute for Space Studies, James Hansen, warns. In a decade, it will be too late to arrest climate change. That's why a grassroots focus is needed. We need action now. Promoting informed, sustainable, and eco-conscious travel can be instrumental here. As American writer and humorist Mark Twain asserted over 150 years ago, "Travel is fatal to prejudice, bigotry, and narrow-mindedness, and many of our people need it sorely on these accounts. Broad, wholesome, charitable views of men and things cannot be acquired by vegetating in one little corner of the earth all one's lifetime."[9]

We're not the only ones needing to change how we interact with the planet, of course. China surpassed us as the leading annual greenhouse gas emitter years ago, in 2006. But we're still among the leaders in per capita admissions and number one in historical emissions for those keeping score at home. What we do matters—a lot. Americans remain the world's heaviest energy users, consuming 15 percent of the world's energy production and almost 20 percent of its electricity, despite constituting only 4 percent of global population. If everyone adopted an American lifestyle, global carbon dioxide emissions would be more than four times what they are today.[10] If everyone lived like Americans, we'd need three additional planets.

Remarkable opportunity exists, paradoxically, given our position as the second leading annual greenhouse gas emitter. *New York Times* columnist and best-selling author Thomas Friedman, for one, asserts America can "get its groove back" by solving the world's biggest problem.[11] As the second half of this book demonstrates, we have already begun this process at local levels across our country. Seven key areas here range from investing in solar and wind energy to alternatives like geothermal, wave, and tidal; from green building and design to urban planning that emphasizes walkability; from reducing consumption to recognizing both the merits and limits of adaptation. While Washington never ratified the original 1997 Kyoto Protocol on climate change and former President Trump removed us from the 2015 Paris Accord, more than nine hundred American cities pledged to meet those updated targets before President Biden brought us back into the accord in 2021. We can still be that shining city on a hill if we make our cities green, as Friedman argues. "A truly green America," he explains, "would be more valuable than fifty Kyoto protocols. Emulation is always more effective than compulsion."[12] Along the way, we can strengthen our civil society by weakening political polarization.

Given such stakes, why so little action?

American writer and political activist Upton Sinclair once wryly stated that it is difficult to see the roots of a problem when one's salary depends on not seeing it. Offering additional detail, University of Victoria professor of psychology and environmental studies Robert Gifford outlines seven psychological barriers that he calls dragons of inaction.[13] Limited cognition leads his list. Think back to the information deficit model introduced early in this book. A global average of 3.6° Fahrenheit change in temperature seems insignificant to most of us. But it's not. That amount is widely considered the point of no return, the tipping point after which any human efforts to address climate change will be to no avail. And that number is more of a political compromise, with 1.5 degrees Celsius (2.7 degrees Fahrenheit), or better yet, 350 to 450 ppm concentration of carbon dioxide in the atmosphere the preferred measuring stick within the scientific community.

But it is the last three "dragons" on Gifford's list that resonate most in my mind.[14] Disbelief of experts, from distrust to outright denial, is one, as the United States experienced with horrifying consequences when it came to use of social distancing, face masks, and vaccinations during the COVID-19 pandemic. Perceived risks of change, from financial to social to psychological, is another, even as distinguished London School of Economics professor and former chief economist of the World Bank Nicholas Stern warns of a procrastination penalty, that later costs will be far greater than those paid upfront. And finally, there is concern about positive but inadequate behavior changes. Scholars like Ernest J. Yanarella and Richard S. Levine call this picking the low-lying fruit.[15] Others name it the rebound effect. The worry is people will become satisfied with small half-measures and fail to address the root of our problem. Even worse, people may reinvest money saved thanks to energy efficiency improvements into greater use of a product. Studies show vehicular fuel-efficiency improvements historically encourage more driving, for example, minimizing overall reductions in greenhouse gas emissions thanks to efficiency enhancements.

Still others suggest inaction on our climate crisis stems from poor framing. Tone is key, these scholars contend. Media predominantly ply doomsday scenarios, rarely countering with inspiring suggestions to curb climate change. Human nature leads us to tune out these negative messages, whereas more positive statements better promote self-efficacy.[16] That's what the second half of this book addresses. Reframing in terms of why we gain from action rather than what we lose from inaction is more emotionally engaging. Hope is more motivating than fear.[17] Beyond tone, framing can position mitigation policies as economic opportunities that create green jobs. It can target technological innovation as a tool reducing local pollutants and providing health benefits. And framing might position climate change as a pressing national security threat, a religious or moral imperative, or a community-building enterprise.[18]

That said, keep in mind, people seek out frames that reinforce their preexisting beliefs. "Existing research shows that people usually select information lining up with prior beliefs and attitudes," academics Thomas Bernauer and Liam McGrath write, "to preserve their existing worldviews, self-concept and self-worth, or to sustain beliefs that are in line with prevailing values, ideologies and beliefs in their social network."[19] That's where new experiences might shake things up, where conscientious travel could foster framing shifts that generate greater climate policy support.

And thankfully, public opinion may not be as ill-informed as we often assume. Begun in 2008 by the Yale Program on Climate Change Communication and George Mason University Center for Climate Change Communication, Global Warming's Six Americas categorizes our nation's opinion on climate change as either alarmed, concerned, cautious, disengaged, doubtful, or dismissive. Notably, an understanding that climate change is happening has increased steadily since 2015. That said, Americans still rank climate change lower in priority than the economy, health care, or national security.

It's not that Americans don't care. We just don't care enough.

That level of significance or degree of importance is what political scientists call salience. When salience combines with intensity of opinion, how strongly those thoughts are held, governments, in the concise words of esteemed political scientist V. O. Key, "find it prudent to heed."[20] Think gun rights advocates or Medicare recipients. To build a broader political base for climate mitigation, climate change concern must become more salient and intense. This happens when an issue is perceived to effect self-interest, namely affecting jobs, income, taxes, war, or health. Marine biologist and author Rachel Carson's seminal *Silent Spring* elevated concern about insecticides in the 1960s when she did precisely this in terms of human health. She made it personal back then. When it comes to climate change, we must, too.

There's a lot to work with here. Americans, according to Gallup and Pew national surveys, increasingly consider climate change at least a somewhat serious problem. Gallup began asking this question in late 1997: Do you think global warming will pose a serious threat to you or your way of life in your lifetime? That first year the response was 25 percent. By 2021, it had risen to 43 percent. That's progress. Admittedly, an expected decline after the Great Recession in 2008 dropped numbers as low as 32 percent in 2010, and attribution of causes has become more contentious. Plus, crucially, most Americans still believe climate change will affect other nations more than our own.

That's why stories of climate change in the United States play such an important role, particularly if they encourage people to see for themselves, to see climate change in action.[21] We can't wait until everyone is on board, though. That train will never leave the station. More would be great, over 50 percent a huge plus, but the challenge today is about motivating the 43 percent already expressing concern. It's not how big that number is. It's about how passionate those people are. Looking at hundreds of campaigns over the past century, Harvard political scientist Erica Chenowith asserts that peaceful mass movements succeed with only 3.5 percent of a population mobilized.[22]

These statistics aside, one thing is certain. Success requires better-informed local knowledge. It requires more Americans understanding how global phenomena impact local communities—and how local communities shape global forces such as climate change. Much maligned international climate efforts are not without merit. Conferences of Parties on climate change have met every year since 1995, aside from 2020 because of the COVID-19 pandemic. Perhaps most notably, after Copenhagen in 2009, negotiators turned away from the top-down apparatus employed in 1997 with Kyoto to a more bottom-up structure. No universal treaty yet exists, even with the 2015 Paris Accord, but we do have a climate change regime complex, as political scientists Robert Keohane and David Victor assert.[23]

That allows more focused, decentralized activities to evolve. Here in the United States, initiatives like Climate Mayors, a bipartisan network representing more than 470 mayors working together to address climate mitigation, is one example. Founded in 2014, it represents 74 million Americans. One of those is sustainability director for the city of Orlando, Chris Castro. "We're still in!" he proudly exclaims about the 2015 Paris Agreement. "The local level is where the rubber meets the road, regardless of what the federal government says."[24] Michael Brune, executive director of the Sierra Club, our oldest and largest environmental organization, agrees that local activity is critically important. "A lot of the work in climate change is local," says Brune. "Every climate change project is in a place, a coal terminal here or pipeline going through there. So many decisions are made locally."[25]

Those local efforts then scale up, Brune adds. "It takes everything. Grassroots isn't solely locally directed," he explains. "Most people, who take action locally, it's not the only thing they do. They also vote, call their representatives, or meet them and advocate for strong policy. Local activism helps to feed and build the strength to global solutions."[26] Emphasizing local action also directs conversation away from political partisanship, as Breakthrough Strategies & Solutions, a climate communications consultancy, asserts.[27] It emphasizes what we have in common and helps rebuild our frayed civil society. That is the playbook for the next decade of climate activism. A combination of witnessing impacts of climate change yourself and prioritizing local actions is the best roadmap to solving our climate crisis. That best builds political salience and efficacy.

Precise policy prescriptions will vary, depending on where folks live, but the seven categories of locally based initiatives detailed over the last half of this book will be constant. The common link throughout will be whether individuals like you advocate those climate solutions at the local level. Join a solar co-op. Connect with an environmental group like Earthwatch or Citizens' Climate Lobby. Talk about climate change with friends and family as well as decision makers in your community, from school administrators to office managers, from those who oversee the restaurants, gyms, theaters, and places of worship you frequent to fellow attendees at dog parks and youth soccer practices. This land depends on it.

NOTES

INTRODUCTION

1. Paul Hond, "The Ice Detectives," *Columbia Magazine*, Fall 2017, 14.
2. Bill McKibben, "The Nature of Crisis," *The New Yorker*, March 26, 2020, https://www.newyorker.com/news/annals-of-a-warming-planet/the-nature-of-crisis-coronavirus-climate-change.
3. Bill McKibben, *Eaarth: Making a Life on a Tough New Planet* (New York: Times Books, 2010), 23.
4. Shannon Osaka, "'The Planet Is Broken,' UN Chief Says," *Grist*, December 3, 2020, https://grist.org/climate/the-planet-is-broken-u-n-leader-says-in-state-of-the-climate-report/.
5. Michael Brune, personal interview, Oakland, CA, July 8, 2021.
6. Jimmy Carter, "Crisis of Confidence" (speech, Washington, DC, July 15, 1979), American Rhetoric, https://www.americanrhetoric.com/speeches/jimmycartercrisisofconfidence.htm.
7. Naomi Klein, *This Changes Everything: Capitalism vs the Climate* (New York: Simon & Schuster, 2014), 117.
8. Paul Hawken, "Beck Environmental Lecture," Fall for the Book, George Mason University, Fairfax, VA, October 10, 2018.
9. Naomi Oreskes and Erik M. Conway, *Merchants of Doubt: How a Handful of Scientists Obscured the Truth on Issues from Tobacco Smoke to Global Warming* (New York: Bloomsbury Press, 2010).
10. Phillip Stoddard, personal interview, April 14, 2021.

11. David Ockwell, Lorraine Whitmarsh, and Saffron O'Neill, "Reorienting Climate Change Communication for Effective Mitigation: Forcing People to Be Green or Fostering Grass-Roots Engagement?" *Science Communication*, January 7, 2009; Abel Gustafson et al., "Personal Stories Can Shift Climate Change Beliefs and Risk Perceptions: The Mediating Role of Emotion," *Communication Reports* 33, no. 3 (August 2020): 121–35.
12. Aldo Leopold, *A Sand County Almanac, and Sketches Here and There* (New York: Oxford University Press, 1949).
13. Akiko Busch, *The Incidental Steward: Reflections on Citizen Science* (New Haven: Yale University Press, 2013), 72.
14. Stoddard interview.
15. Cary Funk and Brian Kennedy, "How Americans See Climate Change and the Environment in 7 Charts," Pew Research Center, April 21, 2020.
16. Richard Louv, *Last Child in the Woods: Saving Our Children from Nature-Deficit Disorder* (Chapel Hill, NC: Algonquin Books, 2005).
17. David W. Orr, *Down to the Wire: Confronting Climate Collapse* (Oxford: Oxford University Press, 2009), 211.
18. Thomas L. Friedman, *Hot, Flat, and Crowded: Why We Need a Green Revolution—and How It Can Renew America* (New York: Farrar, Straus and Giroux, 2008), 32.
19. Orr, *Down to the Wire*, xv.
20. Rice Doyle, "Fla. Gov. Bans the Terms 'Climate Change,' 'Global Warming,'" *USA Today*, March 9, 2015.
21. To be fair, Florida did pass its first ever climate related law in summer 2020, although this legislation is careful to use the term *sea level*, not *climate change*. Still, SB 178 is a start and prohibits spending tax dollars on projects in coastal zones unless studies show that land is safe from rising sea levels. Public Financing of Construction Projects, 2020 Florida Legislature SB 178, Section 161.551, Florida Statutes, effective July 1, 2021, https://flsenate.gov/Session/Bill/2020/178/BillText/er/HTML.
22. Bob Inglis, telephone interview, November 5, 2018.
23. Inglis interview.

1. OUR RISING SEAS

1. Skip Stiles, personal interview, August 17, 2020.
2. David Wallace-Wells, *The Uninhabitable Earth: Life After Warming* (New York: Tim Duggan Books, 2019), 131.
3. Intergovernmental Panel on Climate Change, "Climate Change 2014 Synthesis Report Summary for Policymakers," Section 2.4, https://www.ipcc.ch/report/ar5/syr/.
4. Stiles interview.
5. James A. Baker III, George P. Shultz, and Ted Halstead, "The Strategic Case for U.S. Climate Leadership: How Americans Can Win with a Pro-Market Solution," *Foreign Affairs* 99, no. 3 (May/June 2020): 30.
6. It should be noted that sea level rise due to climate change is not the only culprit. Subsidence connected to the oil industry as well as attempts to control the Mississippi River must also shoulder some share of blame.
7. Rick Van Noy, *Sudden Spring: Stories of Adaptation in a Climate-Changed South* (Athens: University of Georgia Press, 2019), 6.
8. Simone Fiaschi and Shimon Wdowinski, "Local Land Subsidence in Miami Beach (FL) and Norfolk (VA) and Its Contribution to Flooding Hazard in Coastal Communities Along the U.S. Atlantic Coast," *Ocean & Coastal Management* 187 (April 1, 2020), https://doi.org/10.1016/j.ocecoaman.2019.105078.
9. L. Caesar, G. D. McCarthy, D. J. R. Thornalley, et al., "Current Atlantic Meridional Overturning Circulation Weakest in Last Millennium," *Nature Geoscience* 14 (February 25, 2021): 118–20, https://doi.org/10.1038/s41561-021-00699-z.
10. NASA, "Understanding Sea Level—Thermal Expansion," https://sealevel.nasa.gov/understanding-sea-level/global-sea-level/thermal-expansion.
11. The difference in global average temperature between the middle of an ice age and our current climate is less than 9° Fahrenheit.
12. Milankovitch cycles contain three components: eccentricity, obliquity, and precession. Eccentricity refers to the shape of Earth's orbit. Obliquity is the angle Earth's axis is tilted with respect to our orbital plane.

And precession is the direction Earth's axis of rotation is pointed. Together, they affect how much sunlight Earth absorbs from the Sun.

13. In most areas, among them Canada, Scandinavia, New England, and the upper Midwest, that ice melted away ten thousand years ago.
14. Orrin H. Pilkey and Rob Young, *The Rising Sea* (Washington, DC: Island Press, 2009), 79.
15. Pankaj Khanna, André W. Droxler, Jeffrey A. Nittrouer, et al., "Coralgal Reef Morphology Records Punctuated Sea-Level Rise During the Last Deglaciation," *Nature Communications* 19, no. 1 (October 19, 2017): 1046, 10.1038/s41467-017-00966-x.
16. Andrew Guzman, *Overheated: The Human Cost of Climate Change* (Oxford: Oxford University Press, 2013), 84.
17. Union of Concerned Scientists, "Sea Level Rise and Tidal Flooding in Norfolk, Virginia," March 30, 2016, https://ucsusa.org/resources/sea-level-rise-and-tidal-flooding-norfolk-virginia.
18. David Goodrich, *A Hole in the Wind: A Climate Scientist's Bicycle Journey Across the United States* (New York: Pegasus Books, 2017), 19.
19. Union of Concerned Scientists, "Sea Level Rise and Tidal Flooding in Norfolk," March 30, 2016.
20. Katharine Mach, telephone interview, April 16, 2021.
21. Louisiana recognizes two here, the United Houma Nation tribe and the Biloxi-Chitimacha Confederation of Muskogees Inc.
22. Louisiana Office of Community Development, "Resettlement of Isle de Jean Charles: Background & Overview," June 9, 2020, https://isledejeancharles.la.gov/sites/default/files/public/IDJC-Background-and-Overview-1-28-21.pdf.
23. Cory Schouten, "'Climate Gentrification' Could Add Value to Elevation in Real Estate," *MoneyWatch*, December 28, 2017, https://www.cbsnews.com/news/climate-gentrification-home-values-rising-sea-level/.
24. Southeast Florida Regional Climate Change Compact Sea Level Rise Work Group (Compact), February 2020, A document prepared for the Southeast Florida Regional Climate Change Compact Climate Leadership Committee, https://southeastfloridaclimatecompact.org/wp-content/uploads/2020/04/Sea-Level-Rise-Projection-Guidance-Report_FINAL_02212020.pdf.

25. Stan Cox and Paul Cox, "A Rising Tide: Miami Is Sinking Beneath the Sea—But Not Without a Fight," *The New Republic*, November 8, 2015, https://newrepublic.com/article/123216/miami-sinking-beneath-sea-not-without-fight.
26. Kevin Loria, "Miami Is Racing Against Time to Keep Up with Sea-Level Rise," *Business Insider*, April 12, 2018, https://www.businessinsider.com/miami-floods-sea-level-rise-solutions-2018-4.
27. Jim Cason, telephone interview, April 15, 2021.
28. Kevin Loria, "Miami Is Racing."
29. National Oceanic and Atmospheric Administration, "What Is a Perigean Spring Tide?" https://oceanservice.noaa.gov/facts/perigean-spring-tide.html.
30. Jeff Goodell, "Goodbye, Miami," *Rolling Stone*, June 20, 2013, https://www.rollingstone.com/feature/miami-how-rising-sea-levels-endanger-south-florida-200956/.
31. Oliver Milman, "Atlantic City and Miami Beach: Two Takes on Tackling the Rising Waters," *The Guardian*, March 20, 2017, https://www.theguardian.com/us-news/2017/mar/20/atlantic-city-miami-beach-sea-level-rise.
32. David Kamp, "Can Miami Beach Survive Global Warming?" *Vanity Fair*, November 10, 2015, https://www.vanityfair.com/news/2015/11/miami-beach-global-warming.
33. "Politico 50: Our Guide to the Thinkers, Doers and Visionaries Transforming American Politics in 2016," *Politico*, https://www.politico.com/magazine/politico50/2016/philip-stoddard-harold-wanless/.
34. Harold Wanless, personal interview, April 12, 2021.
35. Laura Parker, "Hurricane Matthew's Destructive Storm Surges Hint at New Normal," *National Geographic*, October 8, 2016, https://www.nationalgeographic.com/science/article/hurricane-matthew-storm-surges-predict-sea-level-rise-btf.
36. Wanless interview.
37. Cornelia Dean, *Against the Tide: The Battle for America's Beaches* (New York: Columbia University Press, 1999), 34.
38. James Hansen, Makiko Sato, Paul Hearty, et al., "Ice Melt, Sea Level Rise and Superstorms: Evidence from Paleoclimate Data, Climate Modeling, and Modern Observations that 2°C Global Warming Could Be Dangerous," *Atmospheric Chemistry and Physics* 16 (March 22, 2016):

3761–3812, https://acp.copernicus.org/articles/16/3761/2016/acp-16-3761-2016.pdf.
39. Wanless interview.
40. Herald R. Wanless, "Sea Levels Are Going to Rise by at Least 20ft. We Can Do Something About It," *The Guardian*, April 13, 2021, https://www.theguardian.com/environment/commentisfree/2021/apr/13/sea-level-rise-climate-emergency-harold-wanless.

2. FLOODING IN THE FORECAST

1. Danny Halden, Zoom interview, May 18, 2021.
2. Reza Marsooli, Ning Lin, Kerry Emmanuel, and Kairui Feng, "Climate Change Exacerbates Hurricane Flood Hazards along US Atlantic and Gulf Coasts in Spatially Varying Patterns," *Nature Communications* 10, no. 3785 (2019), https://doi.org/10.1038/s41467-019-11755-z.
3. Tyler Kelley, *Holding Back the River: The Struggle Against Nature on America's Waterways* (New York: Avid Reader Press, 2021).
4. Chris Mooney, "What We Can Say About the Louisiana Floods and Climate Change," *Washington Post*, August 15, 2016.
5. U.S. National Weather Service, "Summary of Natural Hazard Statistics for 2019 in the United States," June 25, 2020, https://www.weather.gov/media/hazstat/sum19.pdf.
6. Matt Hollon, Zoom interview, May 21, 2021.
7. Tom Ennis, Zoom interview, May 21, 2021.
8. In one of those hours, it rose five feet every fifteen minutes.
9. Robert Hanna, personal interview, May 18, 2021.
10. Laura Gutschke, "Rain Not Going Away This Week for Abilene," *Abilene Reporter-News*, May 19, 2021.
11. Srinivas Valavala, personal interview, May 19, 2021.
12. Jessica Ranck, "Abilene Residents Unhappy with Street Drainage System, City Says There's Not Much They Can Do About It," *Big Country Homepage*, April 28, 2021, https://www.bigcountryhomepage.com/news/main-news/abilene-residents-unhappy-with-drainage-system-city-says-theres-not-much-they-can-do-about-it/.
13. NOAA National Centers for Environmental Information, "Billion-Dollar Weather and Climate Disasters: Mapping," July 9, 2021, https://www.ncdc.noaa.gov/billions/mapping.

14. *Testimony of Michael Grimm, Before the Committee on Science, Space and Technology Subcommittee on Investigations and Oversight Subcommittee on Environment*, 116th Congress, February 27, 2020.
15. Ennis interview.
16. In poorer countries especially, a fatal mix of diseases often circulates following floods, from dysentery and other gastrointestinal problems to malaria, smallpox, measles, typhoid, cholera, and skin diseases like scabies, impetigo, and fungal infections.
17. Pete Spotts, "Record-Breaking Floods Force Engineers to Blow up Mississippi River Levee," *Christian Science Monitor*, May 2, 2011, https://www.csmonitor.com/USA/2011/0502/Record-breaking-floods-force-engineers-to-blow-up-Mississippi-River-levee.
18. Charles Maldonado, "Rollin' Down the River," *City Paper* (May 22–24, 2010), 12.
19. The Cumberland River basin covers 17,000 square miles, flowing 688 miles from southeast Kentucky and winding its way through Middle Tennessee before emptying into the Ohio River in southwestern Kentucky.
20. That dubious honor belongs to a flood in 1926–1927 that reached 56.20 feet. However, because there were no dams then, it's difficult to compare floodwaters from then to those of today.
21. Michael Cain, interview by Robin Robinson, "Flood 2010 Oral Histories," Special Collections, Public Library of Nashville and Davidson County, June 30, 2011.
22. Matt Pylkas, interview by Susannah Gibbons, "Flood 2010 Oral Histories," Special Collections, Public Library of Nashville and Davidson County, August 9, 2011.
23. This was part of Lighthouse Christian School, a six-hundred-student school in Antioch, Tennessee.
24. Nashville Public Library, "Nashville Rising: Our Story of the 2010 Flood," May 1, 2016, http://flood.nashvillepubliclibrary.org/.
25. *Raging Water*, dir. Mark Adams, South Carolina ETV, 2016.
26. Brantley Hargrove, "47 Feet High and Rising," *The Nashville Scene* (April 28–May 4, 2011), 18.
27. Interestingly, in their subsequent incident report, the Corps found the Ohio River, where the Cumberland eventually drains, was nearly at flood stage itself, so there really wasn't anywhere for that water to flow, even if the Corps had lowered its lakes earlier in the week.

28. Matt Pulle and Liz Garrigan, "Up the Creek," *Nashville Scene* (July 8–14, 2010), 12.
29. Jason Rossi, "15 American Cities with the Most Homes in Danger of Flooding," Showbiz Cheat Sheet, December 11, 2018, https://www.cheatsheet.com/culture/american-cities-homes-danger-flooding.html/.
30. Ed Leefeldt and Amy Danise, "FEMA'S Upcoming Changes Could Cause Flood Insurance to Soar at the Shore," *Forbes*, March 18, 2021, https://www.forbes.com/advisor/homeowners-insurance/new-fema-flood-insurance-rates/.
31. Hossein Tabari, "Climate Change Impact on Flood and Extreme Precipitation Increases with Water Availability," *Scientific Reports* 10, no. 13768 (2020), https://doi.org/10.1038.
32. Bioswales soak up water between streets and sidewalks. Irish crossings, the opposite of a speed bump, allow water to drain directly within a street. And green roofs as well as permeable pavement materials for parking offer some relief.
33. Frances Stead Sellers, "Charlotte Bulldozes Against Flooding," *Orlando Sentinel*, December 1, 2019.

3. DROUGHT AND WILDFIRE

1. Julie Johnson, Omar Shaikh Rashad, and Matthias Gafni, "Dixie Fire Explodes Near Paradise, Site of the Devastating 2018 Camp Fire," *San Francisco Chronicle*, July 14, 2021.
2. U.S. Forest Service, *The Rising Cost of Wildfire Operations* (Washington DC: U.S. Department of Agriculture, 2015).
3. Michael McKnight, "Paradise Lost and Found," *Sports Illustrated* (November 4, 2019), 64–71.
4. David Leon Zink, personal interview, July 12, 2021.
5. Melissa Schuster, personal interview, July 13, 2021.
6. Bill Hartley, personal interview, July 10, 2021.
7. Donald G. Criswell, personal interview, July 10, 2021.
8. Joe Tapia, personal interview, July 12, 2021.
9. Tapia interview.
10. Collette Curtis, personal interview, July 12, 2021.

11. Sheep grazing in our national forests is yet another human-related factor in wildfire. Sheep reduce understory grasses that would otherwise fuel more frequent but less intense wildfires.
12. Philip Connors, *Fire Season: Field Notes from a Wilderness Lookout* (New York: HarperCollins, 2011), 149.
13. With an annual economic output of $2.9 trillion, California ranks as the fifth-largest economy in the world, just ahead of the United Kingdom.
14. Coral Davenport and Adam Nagourney, "Fighting Trump on Climate, California Becomes a Global Force," *New York Times*, May 25, 2017.
15. Curtis interview.
16. Volker C. Radeloff et al., "Rapid Growth of the US Wildland-Urban Interface Raises Wildfire Risk," *Proceedings of the National Academy of Sciences of the United States of America* 115, no.13 (March 27, 2018): 3314–19, https://doi.org/10.1073/pnas.1718850115.
17. Criswell interview.
18. Criswell interview.
19. Edward Struzik, *Firestorm: How Wildfire Will Shape Our Future* (Washington, DC: Island Press, 2017), 3.
20. Taylor Lorenz, "Are Gender Reveals Cursed?" *New York Times*, September 10, 2020, https://www.nytimes.com/2020/09/10/style/gender-reveal-parties-cursed.html.
21. John Schwartz and Veronica Penney, "In the West, Lightning Grows as a Cause of Damaging Fires," *New York Times*, October 23, 2020, https://www.nytimes.com/interactive/2020/10/23/climate/west-lightning-wildfires.html.
22. In Northern California, these are called "Diablo winds" after Mount Diablo in the eastern Bay Area.
23. M. D. Flannigan and C. E. Van Wagner, "Climate Change and Wildfire in Canada," *Canadian Journal of Forest Research* 21 (1991): 66–72.
24. John T. Abatzoglou and A. Park Williams, "Impact of Anthropogenic Climate Change on Wildfire Across Western US Forests," *Proceedings of the National Academy of Sciences 113*, no. 42 (October 10, 2016): 11770–75, https://www.pnas.org/content/113/42/11770.
25. National Climate Assessment 2018, 65.
26. Gary Ferguson, *Land on Fire: The New Reality of Wildfire in the West* (Portland, OR: Timber Press, 2017), 24.

27. National Climate Assessment 2018, 77.
28. California's 2006 Global Warming Solutions Act, Assembly Bill 32, is the most ambitious climate legislation on our continent, and among the most aggressive policies globally. It mandates 15 percent carbon emissions below 2005 levels by 2020 and 80 percent below 1990 levels by 2050, with a cap-and-trade system across multiple greenhouse gases and sectors of the economy.
29. Rocky Mountain Research Station, "Frequently Asked Questions About the Mountain Pine Beetle Epidemic," U.S. Forest Service, https://www.fs.usda.gov/rmrs/frequently-asked-questions-about-mountain-pine-beetle-epidemic.
30. Caitlin R. Proctor et al., "Wildfire Caused Widespread Drinking Water Distribution Network Contamination," *Water Science* 2, no. 4 (July/August 2020), https://awwa.onlinelibrary.wiley.com/doi/full/10.1002/aws2.1183.
31. J. L. Thomas et al., "Quantifying Black Carbon Deposition Over the Greenland Ice Sheet from Forest Fires in Canada," *Geophysical Research Letters* 44, no. 15 (August 16, 2017): 7965–74, https://agupubs.onlinelibrary.wiley.com/doi/full/10.1002/2017GL073701.
32. Aaron Smith, "U.S. Drought Drives Up Food Prices Worldwide," *CNN Money*, August 9, 2012, https://money.cnn.com/2012/08/09/news/economy/food-prices-index/index.htm.
33. California Climate & Agriculture Network, "Climate Threats to Agriculture," https://calclimateag.org/climatethreatstoag/.
34. Tapan B. Pathak et al., "Climate Change Trends and Impacts on California Agriculture: A Detailed Review," *Agronomy* 8, no. 3 (February 26, 2018), https://doi.org/10.3390/agronomy8030025.
35. Austin Rempel, "Replanting Paradise," *American Forests*, Fall 2020, https://www.americanforests.org/magazine/article/replanting-paradise/.
36. Paige St. John, Joseph Serna, and Rong-Gon Lin II, "Must Reads: Here's How Paradise Ignored Warnings and Became a Deathtrap," *Los Angeles Times*, December 30, 2018, https://www.latimes.com/local/california/la-me-camp-fire-deathtrap-20181230-story.html.
37. "Paradise Nature-Based Fire Resilience Project Final Report," *Paradise Recreation and Park District*, June 2020.
38. Dan Efseaff, telephone interview, July 26, 2021.

39. Some 69 percent of the state believes climate change is increasing the severity of wildfires out west, whereas only 52 percent of Americans overall do. Jennifer Marlon and Abigail Cheskis, "Wildfires and Climate Are Related—Are Americans Connecting the Dots?" Yale Program on Climate Change Communication, Blog, December 11, 2017, https://climatecommunication.yale.edu/news-events/connecting-wildfires-with-climate/.

4. MORE EXTREME WEATHER

1. Joe Mario Pedersen, "Central Florida Hit by EF-2 Tornado with 115 mph Winds, NWS Reports," *Orlando Sentinel*, August 20, 2020.
2. Thomas J. Fox, *Green Town U.S.A.: The Handbook for America's Sustainable Future* (Hobart, NY: Hatherleigh, 2013), 15.
3. Bob Dixson, personal interview, April 21, 2021.
4. The Enhanced Fujita Scale replaced the original Fujita Scale in February 2007 and measures tornadoes according to the strongest winds they pack, more closely aligning wind speeds with damage they cause. It is named after Ted Fujita, a meteorologist who came to the United States from his native Japan in the early 1950s to work at the University of Chicago.
5. Matthew Cappucci, "Tornado Alley in the Plains Is an Outdated Concept. The South Is Even More Vulnerable, Research Shows," *Washington Post*, May 16, 2020, https://www.washingtonpost.com/weather/2020/05/16/tornado-alley-flawed-concept/.
6. Even though it is warm in July and August, tornadoes are less common then because the air is dried out.
7. Staci Derstein, personal interview, April 20, 2021.
8. Shawn Cannon, personal interview, April 23, 2021.
9. Dixson interview.
10. Derstein interview.
11. Matt Christenson, personal interview, April 21, 2021.
12. Hurricanes Charley, Frances, and Jeanne.
13. Jeff Kunerth, "10 Years After Charley, Central Florida's Tree Canopy Springs Back," *Orlando Sentinel*, August 12, 2014, https://www.orlandosentinel.com/news/breaking-news/os-hurricane-charley-anniversary-tree-20140812-story.html.

14. Jeff Goodell, *The Water Will Come: Rising Seas, Sinking Cities, and the Remaking of the Civilized World* (New York: Little, Brown and Company, 2017), 181.
15. Michael E. Mann and Tom Toles, *The Madhouse Effect: How Climate Change Denial Is Threatening Our Planet, Destroying Our Politics, and Driving Us Crazy* (New York: Columbia University Press, 2016), 24.
16. Marshall Shephard, "Are We Experiencing a New Normal with Extreme Weather?" Climate Correction, University of Central Florida, Orlando, October 3, 2019.
17. Lovins coined the term "global weirding," subsequently popularized by *New York Times* columnist Thomas Friedman in 2010.
18. Hurricane season for Americans runs from June 1 to November 30, with the worst storms historically falling between August and September.
19. In the South Pacific and Indian Oceans, these storms are called cyclones. In the Western Pacific, they are known as typhoons. In the North Atlantic and Eastern Pacific, they go by the name hurricanes.
20. National Hurricane Center, "Saffir-Simpson Hurricane Wind Scale," National Oceanic and Atmospheric Administration, https://www.nhc.noaa.gov/aboutsshws.php.
21. The three other Category 5 U.S. hurricanes to make landfall in the last century were also only tropical storms roughly three days prior to landfall. These include Hurricane Andrew in South Florida in 1992, Hurricane Camille in Mississippi in August 1969, and the Labor Day Hurricane of 1935 in the Florida Keys.
22. Kerry Emanuel, "Increasing Destructiveness of Tropical Cyclones Over the Past 30 Years," *Nature* 436 (2005): 686–88.
23. Jack E. Davis, *The Gulf: The Making of an American Sea* (New York: Liveright, 2017), 410.
24. Andrew Guzman, *Overheated: The Human Cost of Climate Change* (Oxford: Oxford University Press, 2013), 88.
25. This surpassed the 2005 record season and its twenty-eight storms. Among these were also a record four Category 5 storms, including the infamous Hurricane Katrina in August 2005. Katrina killed more than 1,800 people and left more than 100,000 homeless when it flooded 80 percent of New Orleans and caused $75 billion in damages to the city.

26. P. J. Webster et al., "Changes in Tropical Cyclone Number, Duration and Intensity in a Warming Environment," *Science* 309 (2005): 1844–46.
27. Lin Li and Pinaki Chakraborty, "Slower Decay of Landfalling Hurricanes in a Warming World," *Nature* 587 (2020): 230–234, https://doi.org/10.1038/s41586-020-2867-7.
28. Jesse Nichols and Eve Andrews, "Climate Change Made Hurricane Harvey Wetter. Here's How We Know," *Grist*, March 6, 2018, https://grist.org/article/climate-change-made-hurricane-harvey-wetter-heres-how-we-know.
29. Andrea Thompson, "The Fingerprints of Global Warming on Extreme Weather," *Climate Central*, April 24, 2017.
30. Michael Oppenheimer, "As the World Burns: Climate Change's Dangerous Next Phase," *Foreign Affairs* 99, no. 6 (November/December 2020): 34–40.

5. THE MELT IS ON

1. Walter R. Borneman, *Alaska: Saga of a Bold Land* (New York: HarperCollins, 2003), 331.
2. Mark Adams, *Tip of the Iceberg: My 3,000-Mile Journey Around Wild Alaska, the Last Great American Frontier* (New York: Dutton, 2018), 11–12.
3. Atlantic Richfield announced finding oil and gas at Prudhoe Bay years earlier in June 1968, but work did not officially begin on a new haul road until late April 1974, while construction on the actual pipeline began in March 1975.
4. Mike Dunleavy, "Governor Dunleavy Announces Final Piece of FY20 Budget," Office of Governor Mike Dunleavy, August 19, 2019, https://gov.alaska.gov/newsroom/2019/08/19/budget_pfd/.
5. M. M. Nistor and I. M. Petcu, "Quantitative Analysis of Glaciers Changes from Passage Canal Based on GIS and Satellite Images, South Alaska," *Applied Ecology and Environmental Research* 13, no. 2 (2015): 535–49.
6. Adams, *Tip of the Iceberg*, 202.
7. Adams, *Tip of the Iceberg*, 203.

8. Adams, *Tip of the Iceberg*, 205.
9. Zaz Hollander, "Coast Guard Suspends Search for Man Missing after Propane Blast at Whittier Dock," *Anchorage Daily News*, July 8, 2019, https://www.adn.com/alaska-news/2019/07/08/two-missing-after-explosion-aboard-fishing-vessel-at-whittier-dock/#.
10. Zaz Hollander, "Boat Explosion and Fire Leaves Part of Whittier Dock Unsafe and Hundreds of Pounds of Fish in Limbo," *Anchorage Daily News*, July 11, 2019, https://www.adn.com/alaska-news/2019/07/11/boat-explosion-and-fire-leaves-part-of-whittier-dock-unsafe-and-hundreds-of-pounds-of-fish-in-limbo/.
11. Esau Sinnok, "My World Interrupted," U.S. Department of Energy, December 8, 2015, https://www.doi.gov/blog/my-world-interrupted.
12. Steve Visser and John Newsome, "Alaskan Village Votes to Relocate Over Global Warming," CNN, August 18, 2016, https://www.cnn.com/2016/08/18/us/alaskan-town-votes-to-move/index.html.
13. Richard L. Thoman Jr., telephone interview, July 30, 2019.
14. R. Thoman and J. E. Walsh, *Alaska's Changing Environment: Documenting Alaska's Physical and Biological Changes Through Observations*, International Arctic Research Center, University of Alaska Fairbanks, August 2019, 8.
15. National Climate Assessment 2018, 57.
16. Marco Tedesco and Alberto Flores d'Arcais, *The Hidden Life of Ice: Dispatches from a Disappearing World* (New York: The Experiment, 2020), 47.
17. Tedesco and d'Arcais, *The Hidden Life of Ice*, 77.
18. Thoman and Walsh, *Alaska's Changing Environment*, 11.
19. We know that ice sheets are changing for three main reasons. For one, in some places the melt is flowing twice as fast as twenty years ago. For another, ice sheet elevation has dropped. Finally, as satellite measurements demonstrate, they are losing mass.
20. Justin Gillis and Jugal K. Patel, "Antarctica Sheds Huge Iceberg that Hints at Future Calamity," *New York Times*, July 13, 2017.
21. Larsen C's smaller sibling, Larsen B, collapsed in less than a month in 2002.
22. Thoman and Walsh, *Alaska's Changing Environment*, 11.
23. Larry Hinzman, telephone interview, August 1, 2019.
24. Hinzman interview.

25. Michaeleen Doucleff, "Is There a Ticking Time Bomb Under the Arctic?" *NPR: Morning Edition*, January 24, 2018, https://www.npr.org/sections/goatsandsoda/2018/01/24/575220206/is-there-a-ticking-time-bomb-under-the-arctic.
26. Henry Fountain, "A Factory of Warming at the Top of the World," *New York Times*, August 24, 2017.
27. Fountain, "A Factory of Warming."
28. Doucleff, "Is There a Ticking Time Bomb?"
29. Rose Keller, personal interview, July 16, 2019.
30. Elizabeth Weise, "Climate Change Could Melt Decades Worth of Human Poop at Denali National Park in Alaska," *USA Today*, March 31, 2019, https://www.usatoday.com/story/news/nation/2019/03/31/climate-change-could-soon-melt-years-worth-human-poop-alaska-park/3299522002/.

6. CHANGING HABITATS AND SPECIES DIVERSITY LOSS

1. Maurice Tamman, "The Great Lobster Rush," Reuters, October 30, 2018, https://www.reuters.com/investigates/special-report/ocean-shock-lobster/.
2. Zoeann Murphy and Chris Mooney, "Gone in a Generation: Across America, Climate Change Is Already Disrupting Lives," *Washington Post*, January 29, 2019, https://www.washingtonpost.com/graphics/2019/national/gone-in-a-generation.
3. Christopher White, *The Last Lobster: Boom or Bust for Maine's Greatest Fishery?* (New York: St. Martin's Press, 2018), 5.
4. The population center shifted over those years from Long Beach Island, New Jersey, to Boston, Massachusetts, 172 miles away.
5. One negative here is that some of these teenagers, enticed by the money, never return to school.
6. White, *The Last Lobster*, 16.
7. Murphy and Mooney, "Gone in a Generation."
8. L. Caesar et al., "Observed Fingerprint of a Weakening Atlantic Ocean Overturning Circulation," *Nature* 556 (April 12, 2018): 191–209.
9. Livia Albeck-Ripka, "Climate Change Brought a Lobster Boom. Now It Could Cause a Bust," *New York Times*, June 21, 2018, https://www.nytimes.com/2018/06/21/climate/maine-lobsters.html.

10. Jérôme Sueur, Bernie Krause, and Almo Farina, "Climate Change Is Breaking Earth's Beat," *Trends in Ecology and Evolution* 34, no. 11 (November 2019): 971–973.
11. It became Acadia National Park in January 1929.
12. George B. Dorr, *The Story of Acadia National Park: The Complete Memoir of the Man Who Made It All Possible* (Bar Harbor, ME: Acadia Publishing Company, 1997), 3.
13. Abe Miller-Rushing, "Managing a Changing Acadia National Park," Acadia and Schoodic Education and Research Center, National Park Service, U.S. Department of Interior, June 25, 2018.
14. National Park Service, "Schoodic Peninsula," https://www.nps.gov/acad/planyourvisit/schoodic.htm.
15. Earthwatch, *Climate Change: Sea to Trees at Acadia National Park* (Boston: Earthwatch Institute, 2018).
16. Nicholas Fisichelli, personal interview, June 28, 2018.
17. Earthwatch, *Climate Change.*
18. C. W. Greene et al, "Vascular Flora of the Acadia National Park Region, Maine," *Rhodora* 107 (2005): 117–85.
19. D. S. Chandler et al., "Biodiversity of the Schoodic Peninsula: Results of the Insect and Arachnid Bioblitzes at the Schoodic District of Acadia National Park, Maine," Maine Agricultural and Forest Experiment Station, University of Maine, Orono, 2012.
20. A. J. Miller-Rushing et al., "Bird Migration Times, Climate Change, and Changing Population Sizes," *Global Change Biology* 14 (2008): 1959–72.
21. R. B. Primack and A. J. Miller-Rushing, "Uncovering, Collecting, and Analyzing Records to Investigate the Ecological Impacts of Climate Change: A Template from Thoreau's Concord," *BioScience* 62 (2012): 170–81.
22. Abe Miller-Rushing, "Managing a Changing Acadia National Park," Science Coordinator, Acadia and Schoodic Education and Research Center, National Park Service, U.S. Department of Interior, June 25, 2018.
23. Since 1970, 40 percent of plants and animals—individual beings, not species—have vanished from Earth.
24. Bees, for example, pollinate habitats only within a few miles of farms.

25. Miyo McGinn, "New Study Pinpoints the Places Most at Risk on a Warming Planet," *Grist*, October 17, 2019, https://grist.org/article/new-study-pinpoints-the-places-most-at-risk-on-a-warming-planet.
26. Mark L. Hineline, *Ground Truth: A Guide to Tracking Climate Change at Home* (Chicago: University of Chicago Press, 2018).
27. A friendship gift from Japan in 1912, roughly three thousand cherry trees ring Washington's Tidal Basin and nearby national monuments. Some 1.5 million people visit during National Cherry Blossom Festival each year, but the festival increasingly struggles to coordinate with its namesake. The average bloom date moved up from April 6 to April 1 between 1921 and 2017. Jason Samenow, "Japan's Cherry Blossoms Signal Warmest Climate in More Than 1,000 Years," *Washington Post*, April 4, 2017, https://www.washingtonpost.com/news/capital-weather-gang/wp/2017/04/04/japans-cherry-blossoms-signal-warmest-climate-in-over-1000-years/.
28. Charlotte Albright and Amy Olson, "Study: As Climate Changes, So Will Maple Syrup Production," *Dartmouth News*, October 2, 2019, https://news.dartmouth.edu/news/2019/10/study-climate-changes-so-will-maple-syrup-production.
29. Tom Henry, "Climate Change Called Certain and Most Predictions Are Bad," *Toledo Blade*, October 13, 2008.
30. Joshua J. Lawler et al., "Planning for Climate Change Through Additions to a National Protected Area Network: Implications for Cost and Configuration," *Philosophical Transactions of the Royal Society B*, January 27, 2020, https://royalsocietypublishing.org/doi/10.1098/rstb.2019.0117.
31. Michelle Ma, "Rethinking Land Conservation to Protect Species that Will Need to Move with Climate Change," *UW News*, January 28, 2020, https://www.washington.edu/news/2020/01/28/rethinking-land-conservation-to-protect-species-that-will-need-to-move-with-climate-change/.
32. Thoreau first suggested this oft-repeated phrase at a Concord lecture on "The Wild" in April 1851. An essay published after his death draws upon this. Henry David Thoreau, "Walking," *The Atlantic Monthly* 9, no. 56 (June 1862): 665.

33. Chris Conway, "Yellowstone's Wolves Save Its Aspens," *New York Times*, August 5, 2007, https://www.nytimes.com/2007/08/05/weekinreview/05basic.html.
34. Vicky Kleinman, personal interview, June 27, 2018.
35. Beverly Anderson, personal interview, June 27, 2018.

7. OCEAN TROUBLE

1. Florida Department of Environmental Protection, "Florida's Coral Reefs," August 5, 2020, https://floridadep.gov/rcp/rcp/content/floridas-coral-reefs.
2. Keys are coral islands rather than sand islands.
3. The remaining Florida Reef lies within Dry Tortugas National Park west of the Marquesas Keys, John Pennekamp Coral Reef State Park (the first undersea park in our country), and isolated coral patch reefs north of Biscayne National Park up to Stuart in Martin County.
4. Leatherbacks are not found here at all, while green, logger, and hawksbill are uncommon.
5. *Chasing Coral*, dir. Jeff Orlowski, Netflix, 2017.
6. Danny Wells, personal interview, April 13, 2021.
7. National Park Service, "The Joneses of Porgy Key: Arthur and Lancelot," Biscayne National Park, June 18, 2020, https://www.nps.gov/bisc/learn/historyculture/the-joneses-of-porgy-key-page-3.htm.
8. Lizette Alvarez, "A Florida City That Never Was," *New York Times*, February 8, 2012, https://www.nytimes.com/2012/02/08/us/islandia-a-florida-city-that-never-was.html.
9. Leslie Kemp Poole, *Biscayne National Park: The History of a Unique Park on the "Edge"* (Washington, DC: National Park Service, 2022).
10. National Park Service, "The Birth of Biscayne National Park," Biscayne National Park, February 2, 2017, https://www.nps.gov/bisc/learn/historyculture/the-birth-of-biscayne-national-park.htm.
11. Jack E. Davis. *The Gulf: The Making of an American Sea* (New York: Liveright, 2017), 5.
12. Florida Department of Environmental Protection, "Florida's Coral Reefs," August 5, 2020, https://floridadep.gov/rcp/rcp/content/floridas-coral-reefs.

13. Zoological Society of London, "Coral Reefs Exposed to Imminent Destruction form Climate Change," news release, July 6, 2009.
14. According to the 2018 National Climate Assessment, a coral bleaching event in western Hawaii in 2015 killed 50 percent of its coral cover. We already noted that even the world's largest barrier reef, the Great Barrier Reef in Australia, approximating the entire U.S. eastern coast, has lost over half its coral.
15. Katharine Q. Seelye, "Ruth Gates, a Champion of Coral Reefs in a Time of Their Decline, Dies at 56," *New York Times*, November 7, 2018.
16. Elaina Hancock, "UConn Research: More Carbon in the Ocean Can Lead to Smaller Fish," *UConn Today*, August 4, 2020, https://today.uconn.edu/2020/08/uconn-research-carbon-ocean-can-lead-smaller-fish/#.
17. The consensus estimate is that pH will reach 7.8 by century's end.
18. The ocean absorbed about 30 percent of emitted anthropogenic carbon dioxide from 1750 to 2011, with 40 percent remaining in the atmosphere and the rest stored on land in plants and soils.
19. Bill McKibben, *Eaarth: Making a Life on a Tough New Planet* (New York: Times Books, 2010), 10.
20. Christopher S. Murray and Hannes Baumann, "Are Long-Term Growth Responses to Elevated pCO_2 Sex-Specific in Fish?" *PLoS One* 15, no. 7 (July 17, 2020), https://journals.plos.org/plosone/article?id=10.1371/journal.pone.0235817.
21. Ove Hoegh-Guldberg, "Ove Hoegh-Guldberg: 'We Can Save the Great Barrier Reef,'" Great Barrier Reef Foundation Blog, April 22, 2021, https://www.barrierreef.org/news/blog/ove-hoegh-guldberg-we-can-save-the-great-barrier-reef.

8. HEAT AND HEALTH

1. Laurel Wamsley, "It Was A Balmy 90 Degrees In Anchorage—For the 1st Time on Record," *NPR*, July 5, 2019, https://www.npr.org/2019/07/05/738905306/it-was-a-balmy-90-degrees-yesterday-in-anchorage-for-the-first-time-on-record.
2. Todd Sanford, Regina Wang, and Alyson Kenward, "The Age of Alaskan Wildfires," *Climate Central*, 2015, http://assets.climatecentral.org/pdfs/AgeofAlaskanWildfires.pdf.

3. Noah S. Diffenbaugh, Deepti Singh, Justin S. Mankin, Daniel E. Horton, Daniel L. Swain, Danielle Touma, Allison Charland, et al., "Quantifying the Influence of Global Warming on Unprecedented Extreme Climate Events," *Proceedings of the National Academy of Sciences* 114, no. 19 (May 9, 2017): 4881–86, https://www.pnas.org/content/114/19/4881.
4. Tara Law, "About 2.5 Million Acres in Alaska Have Burned. The State's Wildfire Seasons Are Getting Worse, Experts Say," *Time*, August 20, 2019, https://time.com/5657188/alaska-fires-long-climate-change/.
5. Summary for Policymakers 3.4, 20.
6. Bill McKibben, "The Climate Crisis," *The New Yorker*, November 4, 2020.
7. A 1995 heat wave killed 739 in Chicago alone, for example.
8. Kate R. Weinberger, Daniel Harris, Keith R. Spangler, et al., "Estimating the Number of Excess Deaths Attributable to Heat in 297 United States Counties," *Environmental Epidemiology* 4, no. 3 (June 2020): 96, https://journals.lww.com/environepidem/Fulltext/2020/06000/Estimating_the_number_of_excess_deaths.1.aspx?context=LatestArticles.
9. Two other categories were "Still Desirable" (colored aqua) and "Definitely Declining" (colored yellow).
10. Robert Nelson, personal interview, April 30, 2021.
11. Kenneth T. Jackson, *Crabgrass Frontier: The Suburbanization of the United States* (New York: Oxford University Press, 1985).
12. Robert K. Nelson, LaDale Winling, Richard Marciano, Nathan Connolly, et al., "Mapping Inequality," in *American Panorama*, ed. Robert K. Nelson and Edward L. Ayers, https://dsl.richmond.edu/panorama/redlining/.
13. Maria Godoy, "In U.S. Cities, The Health Effects of Past Housing Discrimination Are Plain to See," *NPR*, November 19, 2020, https://www.npr.org/sections/health-shots/2020/11/19/911909187/in-u-s-cities-the-health-effects-of-past-housing-discrimination-are-plain-to-see.
14. Jeremy Hoffman, telephone interview, May 3, 2021.
15. J. S. Hoffman, V. Shandas, and N. Pendleton, "The Effects of Historical Housing Policies on Resident Exposure to Intra-Urban Heat: A

Study of 108 US Urban Areas," *Climate* 8, no. 1 (2020): 12, https://doi.org/10.3390/cli8010012.

16. Katie Patrick, "Urban Heat Islands: The Secret Killer You've Never Heard Of, with Jeremy Hoffman, Ph.D.," *How to Save the World Podcast*, Hello World Labs, San Francisco, April 30, 2018.
17. Jeremy Hoffman, "Throwing Shade in RVA," http://jeremyscotthoffman.com/throwing-shade.
18. A heat wave is defined as a period of two or more days where temperatures stay above historical averages.
19. As an aside, the record for the hottest temperature ever recorded on our planet, 134 degrees in July 1913, rests inland, and much further south in Death Valley, California. That number, though, is questionable, with some climatologists recognizing Death Valley's 129-degree mark instead, one set in June 2013 and subsequently tied by Kuwait in 2016 and Pakistan in 2017. Then, Death Valley rewrote the record book after registering 130 degrees in August 2020 and again in July 2021. Jason Samenow, "Death Valley soars to 130 degrees, potentially Earth's highest temperature since at least 1931," *Washington Post*, August 16, 2020, https://www.washingtonpost.com/weather/2020/08/16/death-valley-heat-record/.
20. Bill McKibben, *The End of Nature* (New York: Random House, 1989), 127.
21. M. Burke, S. Hsiang, S., and E. Migue, "Global Non-Linear Effect of Temperature on Economic Production," *Nature* 527 (2015): 235–39, https://doi.org/10.1038/nature15725.
22. David Wallace-Wells, *The Uninhabitable Earth: Life After Warming* (New York: Tim Duggan Books, 2019), 121.
23. Jonah Engel Bromwich, "Extreme Heat Scorches Southern Arizona," *New York Times*, June 26, 2017.
24. U.S. Energy Information Administration, "Today in Energy," July 23, 2018, https://www.eia.gov/todayinenergy/detail.php?id=36692.
25. Lisa Friedman, "Fixing a Major Piece of the Climate Puzzle," *New York Times*, July 14, 2017.
26. Bigger jets like Boeing 737s and Airbus A320s have higher operating thresholds of 126 and 127 degrees, respectively.

27. Zach Wichter, "Too Hot for Takeoff: Air Travel Buffeted by a Capricious Climate," *New York Times*, June 21, 2017.
28. Jenna Gallegos, "Rising Temperatures Could Bump You from Your Flight. Thanks, Climate Change," *Washington Post*, July 3, 2017, https://www.washingtonpost.com/news/energy-environment/wp/2017/07/03/rising-temperatures-could-bump-you-from-your-flight-thanks-climate-change/.
29. Higher temperatures are also changing jet stream winds. At higher altitudes, the jet stream is becoming more intense, making flights bumpier and affecting travel times. This translates into gains flying east and losses flying west, although the two do not regularly offset one another. Losses should prevail.
30. Alan M. Rhoades, Andrew D. Jones, and Paul A Ulrich, "The Changing Character of the California Sierra Nevada as a Natural Reservoir," *Geophysical Research Letters* 45, no. 23 (December 16, 2018): 13008–19.
31. Deborah Netburn, "Sierra Nevada Snowpack on Track to Shrink Up to 79% by the End of the Century, New Study Finds," *Los Angeles Times*, December 16, 2018, https://www.latimes.com/local/lanow/la-me-ln-sierra-nevada-snowpack-20181216-story.html.
32. Tom Knudson, "Sierra Warming, Later Snow, Earlier Melt," *Sacramento Bee*, December 26, 2008.
33. "Study: Global Warming Could Boost Crop Pests," *Chicago Tribune*, December 16, 2008.
34. There is a bit of good news with this warming: parts of Florida and the Gulf Coast more broadly will become too hot for those blacklegged ticks.
35. Sean K. Smith, *You Can Save the Earth* (Long Island City, NY: Hatherleigh Press, 2008), 23.
36. Crystal Gammon, "Changing Climate Increases West Nile Threat in U.S.," *East Bay Times*, March 20, 2009, https://www.eastbaytimes.com/2009/03/20/changing-climate-increases-west-nile-threat-in-u-s/.
37. Kari Lydersen, "Risk of Disease Rises with Water Temperatures," *Washington Post*, October 20, 2008.
38. Kate R. Weinberger et al., "Estimating the Number of Excess Deaths," 96.

39. Taylor Dahl, "6 Surprising Hearth Attack Triggers," *Sharecare*, September 24, 2018, https://www.sharecare.com/health/heart-attack/slideshow/6-surprising-heart-attack-triggers.
40. Tom Henry, "Climate Change Called Certain and Most Predictions Are Bad," *Toledo Blade*, October 13, 2008.
41. Kate Stein, "Temperatures in Florida Are Rising. For Vulnerable Patients, the Heat Can Be Life-Threatening," Florida Center for Investigative Reporting, November 13, 2018, https://www.wlrn.org/news/2018-11-13/temperatures-in-florida-are-rising-for-vulnerable-patients-the-heat-can-be-life-threatening.
42. Yamiche Alcindor, "In Sweltering South, Climate Change Is Workplace Hazard," *New York Times*, August 4, 2017.

9. HERE COMES THE SUN

1. Heather Rogers, "Current Thinking," *New York Times Magazine*, June 3, 2007, https://www.nytimes.com/2007/06/03/magazine/03wwln-essay-t.html.
2. As an aside, it is worth noting that PV panels themselves are not the only expense to going solar. Soft costs of financing, acquisition, permitting, and installation, typically constitute half the expenses within a rooftop system. These have not fallen at the same rate panels have.
3. Solar Energy Industries Association, "Top 10 Solar States," 2020, https://www.seia.org/research-resources/top-10-solar-states-0.
4. Mary Dipboye, personal interview, August 17, 2017.
5. Alan Brand, telephone interview, November 12, 2019.
6. That's down 90 percent since the 1970s.
7. Fred Krupp and Miriam Horn, *Earth: The Sequel* (New York: Norton, 2018), 29.
8. Florida's Property Assessed Clean Energy (PACE) program recognizes this, providing low-interest financing spread over five to twenty years with a mortgage.
9. Michael Cohen, personal interview, March 1, 2018.
10. Julia Pyper, "Florida Voters Defeat Utility-Backed Solar Amendment," *Green Tech Media*, November 9, 2016, https://www.greentech

media.com/articles/read/florida-voters-defeat-utility-backed-solar-amendment.

11. Scott Thomasson, "Vote Solar & Business Groups Commend Governor Rick Scott for Signing onto Solar," *Vote Solar FL*, June 16, 2017, https://votesolar.org/usa/florida/updates/rick-scott-signs-solar-bill/.
12. The utility industry sought to further stack the deck in its favor by requiring petition gatherers to sign an agreement that they would not carry other related petitions under penalty of $5,000 in damages. Though questions were rightly raised about the legality of such a restriction, some petition gatherers still feared repercussions and dropped the pro-solar petition.
13. David Roberts, "Florida's Outrageously Deceptive Solar Ballot Initiative, Explained. Amendment 1 Is a Utility Scam," *Vox*, November 8, 2016, https://www.vox.com/science-and-health/2016/11/4/13485164/florida-amendment-1-explained.
14. Consumers for Smart Solar also won a legal battle in early 2016 challenging the amendment's wording. In a 4–3 ruling, the Florida Supreme Court said the amendment could go forward, even as one judge in the minority called the amendment "a wolf in sheep's clothing."
15. George Cavros, telephone interview, December 11, 2020.
16. Taylor Kate Brown, "Florida's Amendment 1: A Cautionary Tale for 2018?" BBC News, May 27, 2017, https://www.bbc.com/news/world-us-canada-39258421.
17. Rick Garrity, telephone interview, November 5, 2019.
18. Lynn Nilssen, telephone interview, November 12, 2019.
19. Oscar Vargas, telephone interview, March 6, 2018.
20. Corey Ramsden, personal interview, November 12, 2019.
21. Michael Levi, *The Power Surge: Energy, Opportunity, and the Battle for America's Future* (New York: Oxford University Press, 2013), 17.
22. Alec Tyson and Brian Kennedy, "Two-Thirds of Americans Think Government Should Do More on Climate," Pew Research Center, June 23, 2020, https://www.pewresearch.org/science/2020/06/23/two-thirds-of-americans-think-government-should-do-more-on-climate/.
23. Carli Teproff, "New Homes Will Now Require Solar Panels in South Miami, a First in Florida," *Miami Herald*, July 18, 2017, https://www

.miamiherald.com/news/local/community/miami-dade/south-miami/article162307863.html.

24. Nicholas Kusnetz, "South Miami Approves Solar Roof Rules, Inspired by a Teenager," *Inside Climate News*, July 18, 2017, https://insideclimatenews.org/news/18072017/south-miami-florida-solar-roof-rules-sea-level-rise-teenager-activism/.
25. Bobby Magill, "South Miami Just Made a Huge Rooftop Solar Decision," *Climate Central*, July 20, 2017, https://www.climatecentral.org/news/florida-california-solar-mandate-21631.
26. Heather Smith, "How to Convince the Climate Slacker to Get Serious," *Grist*, April 27, 2016, https://grist.org/climate-energy/how-to-convince-the-climate-slacker-to-get-serious.
27. The Germans and Chinese, similarly, brought down solar panel costs through subsidies, allowing their young industries to grow.
28. John Klewin, telephone interview, August 14, 2017.
29. David Coady, Ian Parry, Nghia-Piotre Le, et al., "Global Fossil Fuel Subsidies Remain Large: An Update Based on Country-Level Estimates," IMF Working Paper No. 19/89 (May 2, 2019), https://www.imf.org/en/Publications/WP/Issues/2019/05/02/Global-Fossil-Fuel-Subsidies-Remain-Large-An-Update-Based-on-Country-Level-Estimates-46509.
30. James Ellsmoor, "United States Spend Ten Times More on Fossil Fuel Subsidies Than Education," *Forbes*, June 15, 2019, https://www.forbes.com/sites/jamesellsmoor/2019/06/15/united-states-spend-ten-times-more-on-fossil-fuel-subsidies-than-education/?sh=36fa834b4473.
31. Joseph Aldy, "Money for Nothing: The Case for Eliminating US Fossil Fuel Subsidies," *Resources*, Spring/Summer 2014, https://www.resourcesmag.org/archives/money-for-nothing-the-case-for-eliminating-us-fossil-fuel-subsidies/.
32. National Conference of State Legislatures, "State Renewable Portfolio Standards and Goals," January 4, 2020, https://www.ncsl.org/research/energy/renewable-portfolio-standards.aspx#nmi.
33. Cavros interview.
34. Tim Dickinson, "The Koch Brothers' Dirty War on Solar Power," *Rolling Stone*, February 11, 2016, https://www.rollingstone.com/politics/politics-news/the-koch-brothers-dirty-war-on-solar-power-193325/.

35. Cavros interview.
36. Ramsden interview.
37. John Copenhaver and Joan Frye, "Solar Homeownership Panel—All about Solar," Solar United Neighbors Webinar, September 29, 2020.
38. Dickinson, "The Koch Brothers' Dirty War."
39. Heaven Campbell, personal interview, November 5, 2019.
40. Ramsden interview.

10. LIVING WITH LESS

1. William E. Brown, *Denali: Symbol of the Alaskan Wild* (Virginia Beach, VA: Donning, 1993), 200.
2. One delightfully Alaskan-flavored alternative exists, the road lottery. Each fall, after most tourists depart, this restricted portion of the road opens to a handful of lottery-winning private autos.
3. This need not always be the case. Take, for example, socially conscious investing or how Tesla has made clean cars cool, a status symbol that reflects wealth along with green conscience.
4. Lester R. Brown, *World on the Edge: How to Prevent Environmental and Economic Collapse* (New York: Norton, 2011), 6.
5. Brown, *World on the Edge*, 7.
6. Julian Simon, "Introduction," in *The State of Humanity*, ed. Julian Simon (Malden, MA: Blackwell, 1995), 27.
7. Daisy Kendrick, *The Climate Is Changing: Why Aren't We?* (London: Piatkus, 2020), 143.
8. Jimmy Carter, "Crisis of Confidence," speech, Washington, DC, July 15, 1979, American Rhetoric, https://www.americanrhetoric.com/speeches/jimmycartercrisisofconfidence.htm.
9. Lester R. Brown, *Eco-Economy: Building an Economy for the Earth* (New York: Norton, 2001), 272.
10. Ann Francis, personal interview, August 1, 2017.
11. United Nations Economic Commission for Europe, "Sustainable Energy: Methane Management," https://unece.org/challenge.
12. We've invented some twenty different types of plastic since 1950. These are typically recognizable by their poly monikers: polyethylene, polypropylene, polystyrene, polyvinyl chloride, and polyester.
13. Brown, *Eco-Economy*, 109.

14. Jan Conway, "Volume of Bottled Water in the U.S. 2010–2019." *Statista*, November 26, 2020, https://www.statista.com/statistics/237832/volume-of-bottled-water-in-the-us/.
15. Brown, *World on the Edge*, 114.
16. Kara Lavender Law, Natalie Starr, Theodore R. Siegler, et al., "The United States' Contribution of Plastic Waste to Land and Ocean," *Science Advances* 6, no. 44 (October 30, 2020), https://advances.sciencemag.org/content/6/44/eabd0288.
17. Kendrick, *The Climate Is Changing*, 55.
18. Kendrick, *The Climate Is Changing*, 85, 59.
19. Kendrick, *The Climate Is Changing*, 44, 59.
20. Evan Symon, "Los Angeles City Council Votes to Start Ban on Single Use Plastic Water Bottles," *California Globe*, November 7, 2019, https://californiaglobe.com/section-2/los-angeles-city-council-votes-to-start-ban-on-single-use-plastic-water-bottles/.
21. Jonathan Safran Foer, *We Are the Weather: Saving the Planet Begins at Breakfast* (New York: Farrar, Straus and Giroux, 2019), 133.
22. Environmental Protection Agency, "Sources of Greenhouse Gas Emissions," https://www.epa.gov/ghgemissions/sources-greenhouse-gas-emissions#:~:text=Human%20activities%20are%20responsible%20for,over%20the%20last%20150%20years.&text=The%20largest%20source%20of%20greenhouse,electricity%2C%20heat%2C%20and%20transportation.
23. Fred Krupp and Miriam Horn, *Earth: The Sequel* (New York: Norton, 2008), 193.
24. Michael Bloomberg and Carl Pope, *Climate of Hope: How Cities, Businesses, and Citizens Can Save the Planet*, (New York: St. Martin's Press, 2017), 154.
25. Bill McKibben, *Eaarth: Making a Life on a Tough New Planet* (New York: Times Books, 2010), 176.
26. Pat Murphy, *Plan C: Community Survival Strategies for Peak Oil* (Gabriola Island, BC: New Society, 2008), 195.
27. Foer, *We Are the Weather*, 83, 79, 171.
28. Hannah Ritchie, "Cars, Planes, Trains: Where Do CO_2 Emissions from Transport Come From?" *Our World in Data*, October 6, 2020, https://ourworldindata.org/co2-emissions-from-transport.
29. As just one measure here, we drive so much that one of every five meals in the United States is eaten in a car. Foer, *We Are the Weather*, 83.

30. The electric motor is more than three times as efficient as an internal combustion engine.
31. Winnie Hu, "In New York, Rush Hour Comes to the Bike Lane," *New York Times*, July 31, 2017.
32. Hu, "In New York," 18.
33. Bill Belleville, *Losing It All to Sprawl: How Progress Ate My Cracker Landscape* (Gainesville: University Press of Florida, 2006), 67, 127.
34. Belleville, *Losing It All to Sprawl*, 98.
35. Belleville, *Losing It All to Sprawl*, 98, 100.
36. Paul Owens, personal interview, November 12, 2019.
37. Owens interview.
38. Owens interview.
39. All but $8.5 billion of that was spent by companies instead of our government.
40. Farhad Manjoo, "How Y2K Offers a Lesson for Fighting Climate Change," *New York Times*, July 19, 2017, https://www.nytimes.com/2017/07/19/technology/y2k-lesson-climate-change.html.
41. John Phillimore and Aidan Davison, "A Precautionary Tale: Y2K and the Politics of Foresight," *Futures* 34 (2002): 147–57.
42. Risa Palm, Toby Bolsen, and Justin T. Kingsland, " 'Don't Tell Me What to Do:' Resistance to Climate Change Messages Suggesting Behavior Changes," *Weather, Climate, and Society* 12, no. 4 (2020): 827–835, https://doi.org/10.1175/WCAS-D-19-0141.1.
43. Borrowing from a famous report during the War of 1812, cartoonist Walt Kelly coined the phrase in a 1970 antipollution Earth Day poster before using it again in his Earth Day 1971 comic strip.
44. Orr, *Down to the Wire*, 68.
45. Katharine Hayhoe, *Saving Us: A Climate Scientist's Case for Hope and Healing in a Divided World* (New York: One Signal Publishers/Atria Books, 2021).

11. THE WINDS ARE CHANGING

1. Andrew Myers, "Stanford Study Finds That Wind Energy Output Increases When People Need Heat the Most," *Stanford News*, March 17, 2021, https://news.stanford.edu/2021/03/17/coldest-times-wind-energy-production-heats/.

2. Ken Becker, personal interview, May 19, 2021.
3. Miesha Adames, personal interview, May 19, 2021.
4. U.S. Bureau of Labor Statistics, "Fastest Growing Occupations," *Occupational Outlook Handbook,* April 9, 2021, https://www.bls.gov/ooh/fastest-growing.htm.
5. Elizabeth Weise and Rick Jervis, "As Climate Threat Looms, Texas Republicans Have a Solution: Giant Wind Farm Everywhere," *USA Today*, October 18, 2019, https://www.usatoday.com/story/news/2019/10/18/texas-wind-energy-so-strong-its-beating-out-coal-power/3865995002/.
6. Billie Jones, telephone interview, May 25, 2021.
7. Philip Marcelo, "Wind Turbine Bases Act as Reefs: In Video, Marine Life Thrives in New, Artificial Habitats," *Orlando Sentinel*, February 19, 2018.
8. Penelope Crossley, *Renewable Energy Law: An International Assessment* (Cambridge: Cambridge University Press, 2019), 24.
9. "Top 10 Biggest Wind Farms," *Power Technology.com*, January 20, 2021, https://www.power-technology.com/features/feature-biggest-wind-farms-in-the-world-texas/.
10. Jud Clemente, "The Great Texas Wind Power Boom," *Forbes*, October 11, 2016, https://www.forbes.com/sites/judeclemente/2016/10/11/the-great-texas-wind-power-boom/?sh=25399dc6c6aa.
11. Roger Drouin, "How Conservative Texas Took the Lead in U.S. Wind Power," *Yale Environment 360*, April 9, 2015, https://e360.yale.edu/features/how_conservative_texas_took_the_lead_in_us_wind_power.
12. Weise and Jervis, "As Climate Threat Looms."
13. Kate Galbraith and Asher Price, *The Great Texas Wind Rush: How George Bush, Ann Richards, and a Bunch of Tinkerers Helped the Oil and Gas State Win the Race to Wind Power* (Austin: University of Texas Press, 2013), 3.
14. Weise and Jervis, "As Climate Threat Looms."
15. Galbraith and Price, *The Great Texas Wind Rush*, 109–10.
16. Becker interview.
17. The Panhandle, El Paso, and parts of East Texas are on other U.S. grids.
18. Weise and Jervis, "As Climate Threat Looms."

19. Weise and Jervis, "As Climate Threat Looms."
20. Galbraith and Price, *The Great Texas Wind Rush*, 6.
21. Drouin, "How Conservative Texas Took the Lead."
22. Hydropower and animal-powered mills, sometimes even people-powered, supplied most of the remaining balance.
23. On the negative side, variable-speed turbines require more components, which means higher equipment costs.
24. Installed capacity is how much power is produced when operating at full tilt, 24/7, whereas generation is the amount of power produced over a given time. The ratio of these two numbers is the capacity factor, a measure of how well an installation lives up to its potential.
25. Ryan Wiser and Mark Bolinger, "2018 Wind Technologies Market Report," U.S. Department of Energy, https://www.energy.gov/sites/prod/files/2019/08/f65/2018%20Wind%20Technologies%20Market%20Report%20FINAL.pdf, 37.
26. American Clean Power Association, https://cleanpower.org/.
27. Marianne Rodgers, telephone interview, August 14, 2017.
28. Wallace Erickson, Gregory Johnson, and David Young, "A Summary and Comparison of Bird Mortality from Anthropogenic Causes with an Emphasis on Collisions," in *Bird Conservation Implementation and integration in the Americas: Proceedings of the Third International Partners in Flight Conference*, ed. C. John Ralph and Terrell D. Rich (Albany, CA: U.S. Department of Agriculture, Forest Service, Pacific Southwester Research Station, 2005): 1029–42.
29. Galbraith and Price, *The Great Texas Wind Rush*, 150.
30. John Klewin, telephone interview, August 14, 2017.
31. Matt Christenson, personal interview, April 21, 2021.

12. BUILDING (AND REBUILDING) GREEN

1. Frank Morris, "Kansas Town's Green Dreams Could Save Its Future," *All Things Considered*, December 27, 2007.
2. Matt Christenson, personal interview, April 21, 2021.
3. Thomas J. Fox, *Green Town U.S.A.: The Handbook for America's Sustainable Future* (Hobart, NY: Hatherleigh, 2013), 206–7.
4. Auden Schendler and Randy Udall, "LEED is Broken; Let's Fix It," *Grist*, October 26, 2005.

5. Steve Hewitt, telephone interview, May 7, 2021.
6. Fredric Heeren, "Rebuilding Greensburg Green," *Smithsonianmag.com*, February 27, 2009, https://www.smithsonianmag.com/science-nature/rebuilding-greensburg-green-55848425/.
7. *Greensburg: A Story of Community Rebuilding*, dir. Brian Schodore, Discovery Communications, 2009.
8. Adam Nossiter, "An Empty Place Where a Kansas Town Once Stood," *New York Times*, May 7, 2007.
9. Bob Dixson, personal interview, April 21, 2021.
10. Mark Jaccard, *The Citizen's Guide to Climate Success: Overcoming Myths that Hinder Progress* (Cambridge: Cambridge University Press, 2020), 199.
11. Scott Bitikofer, personal interview, August 2, 2017.
12. Thomas Friedman, *Hot, Flat, and Crowded: Why We Need a Green Revolution—and How It Can Renew America* (New York: Farrar, Straus and Giroux, 2008), 283.
13. Christenson interview.
14. Alex Kopestinsky, "Electric Car Statistics in the US and Abroad," *Policy Advice*, April 6, 2021, https://policyadvice.net/insurance/insights/electric-car-statistics/.
15. Tom Krisher, "US Report: Gas Mileage Down with Emissions Up," *Orlando Sentinel*, January 7, 2021.
16. Paul Owens, personal interview, November 12, 2019.
17. "Electric Vehicles Can Fight Climate Change, But They're Not a Silver Bullet: U of T Study," *University of Toronto News*, October 2, 2020, https://www.utoronto.ca/news/electric-vehicles-can-fight-climate-change-they-re-not-silver-bullet-u-t-study.
18. Niraj Chokshi, "Biden's Push for Electric Cars: $174 Billion, 10 Years and a Bit of Luck," *New York Times*, March 31, 2021.
19. Electric cars are cheaper to own, costing less to power on a per mile basis than gasoline and requiring less routine maintenance.
20. For comparison's sake, there are 115,000 gas stations in the United States, most with multiple pumps.
21. Paul Brooker, personal interview, July 21, 2017.
22. Brooker interview.
23. Environmental Protection Agency, "About Energy Star," https://www.energystar.gov/about.

24. Lester R. Brown, *World on the Edge: How to Prevent Environmental and Economic Collapse* (New York: Norton, 2011), 104.
25. Architecture 2030, "The 2030 Challenge," https://architecture2030.org/2030_challenges/2030-challenge/.
26. Philip M. Donovan, personal interview, September 5, 2019.

13. ADDITIONAL ALTERNATIVE ENERGIES

1. Tatiana Serafin, "Dumbest Business Idea of the Year," *Forbes*, July 5, 2004, https://www.forbes.com/celebrities2004/064a.html.
2. Jeff Richardson, "Chena Hot Springs Resort Owner Bernie Karl Named UAF's Business Leader of the Year," *Fairbanks Daily News-Miner*, March 28, 2010, http://www.newsminer.com/news/local_news/chena-hot-springs-resort-owner-bernie-karl-named-uaf-s/article_db69bfaa-1361-5249-9663-2d701324fa8f.html.
3. U.S. Department of Energy, Office of Energy Efficiency and Renewable Energy, "Geovision: Harnessing the Heat Beneath Our Feet," May 29, 2019, https://www.energy.gov/eere/geothermal/downloads/geovision-harnessing-heat-beneath-our-feet.
4. Jim Robbins, "Can Geothermal Power Play a Key Role in the Energy Transition?" *Yale Environment 360*, December 22, 2020, https://e360.yale.edu/features/can-geothermal-power-play-a-key-role-in-the-energy-transition.
5. Bernie Karl, telephone interview, February 19, 2021.
6. Additional major players include the Alaska Division of Geological & Geophysical Surveys, Republic Geothermal, Alaska Energy Authority, and the University of Alaska.
7. Karl interview.
8. Boise, Idaho, is. With the largest U.S. geothermal heating system, Boise has been using geothermal to heat a third of its downtown since 1983.
9. Ed Maibach, "Climate Change Communication . . . Recent Public Perceptions—Where are We?" EBI Energy & Environment Summit, Law Firm of Hunton, Andrews, Kurth, Washington, DC, November 16, 2018.
10. Jonathan Symons, *Ecomodernism: Technology, Politics, and the Climate Crisis* (Cambridge: Polity Press, 2019).

11. Penelope Crossley, *Renewable Energy Law: An International Assessment* (Cambridge: Cambridge University Press, 2019), 254.
12. Take the example of the Three Gorges Dam Project in China, constructed between 1994 and 2006. It submerged 13 cities, 140 towns, and 1,350 villages, displacing an estimated 1.3 million people.
13. Smaller scale hydropower, considered less than 100 MW, tends to avoid construction of dams, relying on natural river flow, instead.
14. Crossley, *Renewable Energy Law*, 253.
15. Surfing destinations, not surprisingly, are wave-energy hot spots. East–west trade winds typically blow from 30 to 60 degrees latitude, giving the West Coast our greatest wave activity. Key locations for tidal energy, on the other hand, concentrate along the northeastern coast of the United States, as well as the western United Kingdom and the shoreline of South Korea. Small islands are good candidates for both wave and tidal.
16. The first large tidal generating facility was built in France in the early 1970s and still operates today, generating 240 megawatts.
17. David Thill, "Maine Company Looks to Tidal Power as Renewable Energy's Next Generation," Energy News Network, September 23, 2020, https://energynews.us/2020/09/23/northeast/maine-company-looks-to-tidal-power-as-renewable-energys-next-generation/.
18. Fred Krupp and Miriam Horn, *Earth: The Sequel* (New York: Norton, 2008), 129.
19. Biodiesel fueled with palm oil, largely from Southeast Asia, may be even worse as rainforests are typically cut to grow it. Friends of the Earth calculates 87 percent of Malaysian deforestation from 1985 to 2000 stems from new palm oil plantations. This deforestation creates more greenhouse gases than switching to biofuels eliminates.
20. Woody crop examples include shrub willow, eucalyptus, and poplar.
21. Other species include big bluestem, sundial lupine, rigid goldenrod, and tall blazing star.
22. Michael Bloomberg and Carl Pope, *Climate of Hope: How Cities, Businesses, and Citizens Can Save the Planet* (New York: St. Martin's Press, 2017), 77.
23. For context, China produces 80 percent of its electricity from coal, a big part of how they passed us in 2006 to become the world's largest emitter of carbon dioxide.

24. Anthony Ingraffea, "Gangplank to a Warm Future," *New York Times*, July 28, 2013.
25. There's a bit of irony in how public perception of nuclear power evolved. Its global spread dates to the 1953 Atoms for Peace speech President Dwight Eisenhower gave at the United Nations, offering U.S. aid in the form of nuclear energy to any country that swore off nuclear weapons.
26. Paul Hawken, *Drawdown* (New York: Penguin Books, 2017), 20.
27. Only Ecuador legally defines it as such.
28. Nuclear power plants are classified by generation. Generation 1 are the oldest, coming online in the 1950s. They are now almost entirely decommissioned, since the Nuclear Regulatory Commission (NRC) licenses U.S. commercial nuclear reactors for only forty years. Generation 2 is what we find in the United States today. Unlike Generation 1, it uses water instead of graphite to slow nuclear chain reactions. It also uses enriched uranium for fuel instead of natural uranium.
29. Naomi Klein, *This Changes Everything: Capitalism vs the Climate* (New York: Simon & Schuster, 2014), 58.
30. David W. Orr, *Down to the Wire: Confronting Climate Collapse* (Oxford: Oxford University Press, 2009), 29.
31. Lester R. Brown, *Eco-Economy: Building an Economy for the Earth* (New York: Norton, 2001), 98.
32. Bill McKibben, "The Climate Crisis," *The New Yorker*, August 12, 2020.
33. Karl interview.
34. Christopher Hayes, "The New Abolitionism: Averting Planetary Disaster Will Mean Forcing Fossil Fuel Companies to Give Up at Least $10 Trillion in Wealth," *The Nation*, May 12, 2014.

14. RETHINKING OUR CITIES

1. Robert D. Putnam, *Bowling Alone: The Collapse and Revival of American Community* (New York: Simon & Schuster, 2000).
2. Jane Jacobs, *The Death and Life of Great American Cities* (New York: Random House, 1961), 72.
3. Lewis Mumford, *The Culture of Cities* (New York: Harcourt, Brace and Company, 1938).

4. Jeff Speck, *Walkable City: How Downtown Can Save America, One Step at a Time* (New York: Farrar, Straus, and Giroux, 2012), 4.
5. Speck, *Walkable City*, 11.
6. Tim Maslow, personal interview, July 31, 2017.
7. Jacobs, *The Death and Life of Great American Cities*, 65–66.
8. Transport for London, *Town Centre Study*, September 2011, ii.
9. Kelly J. Clifton, Sara Morrisey, and Chloe Ritter, "Business Cycles: Catering to the Bicycling Market," *TR News* (May–June 2012), 29.
10. Dashka Slater, "Walk the Walk," *The New York Times Magazine*, April 20, 2008.
11. Speck, *Walkable City*, 41.
12. Seven of the top ten most dangerous metro areas for walking are in Florida, with Orlando's metro area regularly ranking an unwelcome first in pedestrian deaths.
13. Kevin Spear, "Orlando No.1 Again for Pedestrian Deaths," *Orlando Sentinel*, January 23, 2019, https://www.orlandosentinel.com/news/transportation/os-ne-orlando-deadliest-pedestrians-worsening-20190122-story.html.
14. Matthew E. Kahn, *Green Cities: Urban Growth and the Environment* (Washington, DC: Brookings Institution Press, 2006), 129.
15. Andrés Duany, Elizabeth Plater-Zyberk, and Jeff Speck, *Suburban Nation: The Rise of Sprawl and the Decline of the American Dream* (Berkeley, CA: North Point Press, 2000), 45.
16. Duany, Plater-Zyberk, and Speck, *Suburban Nation*, 116.
17. Mike Houck, personal interview, May 23, 2018.
18. Philip Langdon, "Redeveloping with Pedestrians in Mind," in *Within Walking Distance: Creating Livable Communities for All* (Washington, DC, Island Press, 2017).
19. Margie Boule, "Pearl District's Namesake Was a Jewel of a Woman," *The Oregonian*, April 14, 2002, https://www.oregonlive.com/portland/2002/04/pearl_districts_namesake_was_a.html.
20. Bruce Stephenson, personal interview, June 30, 2021.
21. R. Bruce Stephenson, *Portland's Good Life: Sustainability and Hope in an American City* (Lanham, MD: Lexington Books, 2021).
22. Reza Farhoodi, personal interview, May 25, 2018.
23. Duany, Plater-Zyberk, and Speck, *Suburban Nation*, 199.

24. Craig Ustler, personal interview, March 8, 2018.
25. "The Numbers," *Portland Business Journal*, May 18, 2018, 10.
26. City of Portland, Ordinance 188163. Inclusionary Housing Program Administrative Rule Adopted by City Council. December 21, 2016.
27. Alberto Vargas, personal interview, March 29, 2018.
28. Joint Center for Housing Studies of Harvard University, "The State of the Nation's Housing 2020," https://www.jchs.harvard.edu/state-nations-housing-2020.
29. Seth Wynes and Kimberly A. Nicholas, "The Climate Mitigation Gap: Education and Government Recommendations Miss the Most Effective Individual Actions," *Environmental Research Letters* 12, no. 7 (July 12, 2017), https://web.archive.org/web/20200829162700/https://iopscience.iop.org/article/10.1088/1748-9326/aa7541.
30. Janette Sadik-Khan and Seth Solomonow, *Street Fight: Handbook for an Urban Revolution* (New York: Viking, 2016), 252.
31. Interestingly, one group of Americans is driving less, our younger ones. According to analysis of Federal Highway Administration data by the Green Car Congress, approximately 61 percent of eighteen-year-olds in the United States had a driver's license in 2018, compared to 80.4 percent in 1983.
32. Hope Jahren, *The Story of More: How We Got to Climate Change and Where to Go from Here* (New York: Vintage Books, 2020), 96.
33. Taras Grescoe, *Strap Hanger: Saving Our Cities and Ourselves from the Automobile* (New York: Times Books/Henry Holt and Company, 2012), 14.
34. Lester R. Brown, *World on the Edge: How to Prevent Environmental and Economic Collapse* (New York: Norton, 2011), 61–62.
35. In the words of Andrés Duany, famed American architect and founder of the Congress for the New Urbanism, "Parking is destiny."
36. Adam Brinklow, "It Costs $38,000 to Create One Parking Space in San Francisco," *Curbed SF*, June 8, 2016, https://sf.curbed.com/2016/6/8/11890176/it-costs-38000-to-create-one-parking-space-in-sf.
37. Donald Shoup, *The High Cost of Free Parking* (Chicago: University of Chicago Press, 2008), 189.
38. Sadik-Khan and Solomonow, *Street Fight*, 27.
39. Lester R. Brown, *Eco-Economy: Building an Economy for the Earth* (New York: Norton, 2001), 243.

40. Our interstate highway's origins are military. A young Dwight D. Eisenhower was part of a military convoy that required two months to travel from Washington, DC, to San Francisco in 1919. Years later, as Supreme Commander of the Allied Forces, he saw much different infrastructure in Germany, writing in his memoirs, "Germany had made me see the wisdom of broader ribbons across the land."
41. That's something cities like Portland as well as Milwaukee, San Francisco, Baltimore, and New Haven recognized, and are trying to partially correct, reviving old neighborhoods by removing portions of their inner-city expressways.
42. Jane Holtz Kay, *Asphalt Nation: How the Automobile Took Over America and How We Can Take It Back* (Berkeley: University of California Press, 1998), 129.
43. Brown, *World on the Edge*, 112.
44. Sadik-Khan and Solomonow, *Street Fight*, 235.
45. Matthew E. Kahn, *Green Cities: Urban Growth and the Environment* (Washington, DC: Brookings Institution Press, 2006), 118.
46. Grescoe, *Strap Hanger*, 233.
47. Jeff Mapes, *Pedaling Revolution: How Cyclists Are Changing American Cities* (Corvallis: Oregon State University Press, 2009), 14.
48. Speck, *Walkable City*, 191.
49. Rachel Kaufman, "These Are the Most Bikeable Cities in America," NextCity.org, May 16, 2018, https://nextcity.org/daily/entry/these-are-the-most-bikeable-cities-in-america.
50. Kahn, *Green Cities*, 2.
51. Admittedly, the COVID-19 pandemic raised important public health questions about cities and density, spurring those with means to retreat to smaller towns and the countryside in 2020 and 2021.
52. Michael Bloomberg and Carl Pope, *Climate of Hope: How Cities, Businesses, and Citizens Can Save the Planet* (New York: St. Martin's Press, 2017), 21–22.
53. Sadik-Khan and Solomonow, *Street Fight*, 64.
54. Jacobs, *The Death and Life of Great American Cities*, 448.
55. Jacobs, *The Death and Life of Great American Cities*, 196–97.
56. Richard Florida, *The Rise of the Creative Class, Revisited* (New York: Basic Books, 2011).

15. LIVING WITH CHANGE

1. Punta Gorda, St. Petersburg, and Warm Mineral Springs, Florida, also claim to hold the mythical fountain.
2. Sam Anderson, "Searching for the Fountain of Youth," *New York Times*, October 24, 2014, https://www.nytimes.com/2014/10/26/magazine/my-search-for-the-fountain-of-youth.html.
3. St. Augustine is a popular tourist destination throughout the year, with more than six million visitors annually. The Fourth of July is a big draw, and the three months around the Christmas holidays are another, with the city hosting its famous Nights of Lights festival.
4. Brendan Rivers, "Deluged by Floods, America's 'Oldest City' Struggles to Save Landmarks from Climate Crisis," *The Guardian*, October 28, 2020, https://www.theguardian.com/us-news/2020/oct/28/st-augustine-florida-floods-climate-crisis.
5. Michael Oppenheimer, "As the World Burns: Climate Change's Dangerous Next Phase," *Foreign Affairs* 99, no. 6 (November/December 2020): 36.
6. Intergovernmental Panel on Climate Change, "Climate Change 2014 Synthesis Report Summary for Policymakers," 17, https://www.ipcc.ch/report/ar5/syr/.
7. Intergovernmental Panel on Climate Change, "Climate Change 2014 Synthesis Report," 19.
8. Nancy Fresco, personal interview, July 18, 2019.
9. The National Academy of Sciences defines resilience as the "ability to prepare and plan for, absorb, recover from, and more successfully adapt to adverse events."
10. Southeast Florida Regional Climate Change Compact, "Advancing Resilience Solutions through Regional Action," https://southeastfloridaclimatecompact.org/.
11. Kathleen Dean Moore, *Great Tide Rising: Towards Clarity and Moral Courage in a Time of Planetary Change* (Berkeley: Counterpoint, 2016), 191.
12. "Storm Tide Sweeps City of St. Augustine," *St. Augustine Record*, October 18, 1910.
13. Jean Parker Waterbury, "The Castillo Years: 1668–1763," in *The Oldest City: St. Augustine Saga of Survival*, ed. Jean Parker Waterbury (St. Augustine, FL: St. Augustine Historical Society, 1983), 58.

14. Reuben Franklin, personal interview, August 28, 2017.
15. Sheldon Gardner, "Advocacy Group Part of St. Augustine's Sea Level Rise Efforts," *St. Augustine Record*, March 26, 2017, http://staugustine.com/news/local-news/2017-03-26/advocacy-group-part-st-augustine-s-sea-level-rise-efforts.
16. Leanne Giordono, Hilary Boudet, and Alexander Gard-Murray, "Local Adaptation Policy Responses to Extreme Weather Events," *Policy Sciences* 53 (August 18, 2020): 609–36, https://doi.org/10.1007/s11077-020-09401-3.
17. Magen Wilson, personal interview, July 25, 2017.
18. Franklin interview.
19. Susan Parker, personal interview, July 25, 2017.
20. "St. Augustine's Lost Seawall," *Historic Preservation* 17, no. 4 (July–August 1965), 139.
21. John Regan, personal interview, August 3, 2017.
22. David Kelley, "Symposium: Sub-Tropical and Tropical Coastal Resilience: Social, Economic, and Physical Adaptations in South Florida and the Caribbean," University of Miami, April 7, 2021, https://resilience.miami.edu/index.html.
23. Jon Gertner, "Should the United States Save Tangier Island from Oblivion," *New York Times Magazine*, July 6, 2016, https://www.nytimes.com/2016/07/10/magazine/should-the-united-states-save-tangier-island-from-oblivion.html.
24. Liz Koslov, "The Case for Retreat," *Public Culture* 28, no. 2 (May 1, 2016): 359–87, https://doi.org/10.1215/08992363-3427487.
25. Jim Cason, telephone interview, April 15, 2021.
26. Beyond retreat and seawalls, along with coastal armoring like bulkheads, dikes, and groins, a third adaptation to coastal erosion is beach replenishment. Beginning in 1922 with Coney Island's beaches and continuing along the East Coast and Gulf Coast barrier islands over the last century, more than three hundred beaches in the United States have been "enhanced" in this manner, many multiple times. From Hilton Head and Myrtle Beach, South Carolina, to Ocean City, Maryland, and Miami Beach, Daytona Beach, and Jacksonville Beach in Florida, this involves dredging nearby inlets and bays, then filling the eroding beach with that recovered sand. This is costly, from $1 to $10 million per mile, according to Orrin Pilkey and Rob Young. And it is

temporary, lasting on average less than five years. Orrin H. Pilkey and Rob Young, *The Rising Sea* (Washington, DC: Island Press, 2009), 161.

27. NOAA, "Understanding Living Shorelines," https://www.fisheries.noaa.gov/insight/understanding-living-shorelines.
28. That's a relative bargain compared to the twenty-three families living on Isle de Jean Charles, Louisiana, which lost 98 percent of its land to flooding. In 2016 the U.S. government spent $48 million to resettle them.
29. More of that approach is needed. Some 50 percent the world's wetlands have been lost since 1900. In New York City, over the course of its development, 90 percent of its wetlands have been destroyed. With these losses, water has nowhere else to go.
30. Holdouts necessitate continued municipal services from garbage pickup and water service to streetlights, firefighting, and road maintenance, even as the tax base becomes smaller and smaller to pay for those services. One example is just south of St. Augustine in Summer Haven, Florida. In the 1970s, state officials grew weary of rebuilding SR A1A after every storm and moved the road farther inland. That left over one and a half miles of the old road to the county for maintenance. In 2004, the old A1A washed out again, with the cost to repair it reaching nearly $1 million. By 2008, after yet another washout, residents sued the county for failing to maintain it. Sixty-five property owners collectively alleged that inaction constituted an illegal "taking" under the fifth amendment to the U.S. Constitution. The local court ruled for the county, but a district court overturned that ruling, finding in favor of the property owners. Finally, in 2014, the parties reached a settlement.
31. Naomi Klein, *This Changes Everything: Capitalism vs the Climate* (New York: Simon & Schuster, 2014), 255.
32. Michael E. Mann and Tom Toles, *The Madhouse Effect: How Climate Change Denial Is Threatening Our Planet, Destroying Our Politics, and Driving Us Crazy* (New York: Columbia University Press, 2016), 119.
33. Simon Nicholson, "Solar Radiation Management," Wilson Center, September 30, 2020, https://www.wilsoncenter.org/article/solar-radiation-management#footnote3.
34. This requires a separate geoengineering scheme called ocean fertilization (OF). It involves seeding our seas with iron filings that stimulate

plankton growth. These plankton would soak up carbon via photosynthesis, then drift to the bottom and bury their carbon with them when they die.

35. Gaia Vince, "Sucking CO_2 from the Skies with Artificial Trees," *BBC Future*, October 3, 2012, https://www.bbc.com/future/article/20121004-fake-trees-to-clean-the-skies.
36. Mark Jaccard, *The Citizen's Guide to Climate Success: Overcoming Myths That Hinder Progress* (Cambridge: Cambridge University Press, 2020), 181.
37. Another strategy is to encourage the growth of marshes on inland sides. These grasses trap sediment that otherwise drifts into the water.
38. Nathanael Johnson, "Low on Water, California Farmers Turn to Solar Farming," *Grist*, August 6, 2019, https://grist.org/article/california-farmer-solar-panel-water-renewable-nature-conservancy.
39. Miriam Horn, *Rancher, Farmer, Fisherman: Conservation Heroes of the American Heartland* (New York: Norton, 2016), 129.
40. Susan Glickman, telephone interview, December 2, 2020.
41. Rick Van Noy, *Sudden Spring: Stories of Adaptation in a Climate-Changed South* (Athens: University of Georgia Press, 2019), 6.

CONCLUSION

1. Malcolm Gladwell, *The Tipping Point: How Little Things Can Make a Big Difference* (Boston: Little, Brown & Company, 2000).
2. Riley E. Dunlap, Aaron M. McCright, and Jerrod H. Yarosh, "The Political Divide on Climate Change: Partisan Polarization Widens in the U.S.," *Environment* 58, no. 5 (September 2, 2016): 4–23.
3. Jennifer Marlon, Peter Howe, Matto Mildenberger, et al., "Yale Climate Opinion Maps 2020," Yale Program on Climate Change Communication. September 2, 2020, https://climatecommunication.yale.edu/visualizations-data/ycom-us/.
4. Alexa Spence, Wouter Poortinga, and Nick Pidgeon, "The Psychological Distance of Climate Change," *Risk Analysis* 32, no. 6 (June 1, 2012): 957–972.
5. Ed Maibach, "Climate Change Communication . . . Recent Public Perceptions—Where Are We?" EBI Energy & Environment Summit,

Law Firm of Hunton, Andrews, Kurth, Washington, DC, November 16, 2018.

6. Stuart A. Thompson and Yaryna Serkez, "Every Place Has Its Own Climate Risk. What Is It Where You Live?" *New York Times*, September 18, 2020, https://www.nytimes.com/interactive/2020/09/18/opinion/wildfire-hurricane-climate.html.
7. Steven Levitsky and Daniel Ziblatt, *How Democracies Die* (New York: Crown, 2018).
8. Nathaniel Persily, "Introduction," in *Solutions to Political Polarization in America*, ed. Nathaniel Persily (New York: Cambridge University Press, 2015), 6.
9. Mark Twain, *The Innocents Abroad* (Hartford, CT: American Publishing Company, 1869).
10. Hope Jahren, *The Story of More: How We Got to Climate Change and Where to Go from Here* (New York: Vintage Books, 2020), 83.
11. Thomas Friedman, *Hot, Flat, and Crowded: Why We Need a Green Revolution—and How It Can Renew America* (New York: Farrar, Straus, and Giroux, 2008), 5.
12. Friedman, *Hot, Flat, and Crowded*, 176.
13. Robert Gifford, "The Dragons of Inaction: Psychological Barriers that Limit Climate Change Mitigation and Adaptation," *American Psychologist* 66, no. 4 (May–June 2011): 290–302.
14. Ideological worldview and comparisons with key people follow as the second and third factors for Gifford, while sunk costs combine with behavioral momentum as a fourth category.
15. Ernest J. Yanarella and Richard S. Levine, "Research and Solutions: Don't Pick the Low-Hanging Fruit! Counterintuitive Policy Advice for Achieving Sustainability," *Sustainability: The Journal of Record* 1, no. 4 (August 2008), https://doi.org/10.1089/SUS.2008.9945.
16. Thomas A. Morton, Anna Rabinovich, Dan Marshall, et al., "The Future That May (or May Not) Come: How Framing Changes Responses to Uncertainty in Climate Change Communications," *Global Environmental Change* 21, no. 1 (February 2011): 103–9.
17. Daniel A. Chapman, Brian Lickel, and Ezra M. Markowitz, "Reassessing Emotion in Climate Change Communication," *Nature Climate Change* 7 (November 2017): 850–52.

18. P. S. Hart and E. C. Nisbet, "Boomerang Effects in Science Communication: How Motivated Reasoning and Identity Cues Amplify Opinion Polarization About Climate Mitigation Policies," *Communication Research* 39, no. 6 (2012): 701–23; Dan Kahan, "Fixing the Communications Failure," *Nature* 463, no. 7279 (2010): 296–97; Ezra M. Markowitz and Meaghan L. Guckian, "Climate Change Communication: Challenges, Insights, and Opportunities," in *Psychology and Climate Change*, ed. Susan Clayton and Christie Manning (San Diego, CA: Elsevier, 2018), 35–63; Matthew C. Nisbet, "Communicating Climate Change: Why Frames Matter for Public Engagement," *Environment: Science and Policy for Sustainable Development* 51 (2009): 12–23; Warren Pearce, Brian Brown, Brigitte Nerlich, et al., "Communicating Climate Change: Conduits, Content, and Consensus," *Wiley Interdisciplinary Reviews: Climate Change* 6, no. 6 (2015): 613–26.
19. Thomas Bernauer and Liam F. McGrath, "Simple Reframing Unlikely to Boost Public Support for Climate Policy," *Nature Climate Change* 6, (July 2016): 680–83.
20. V. O. Key, *Public Opinion and American Democracy* (New York: Knopf, 1961), 14.
21. Abel Gustafson, Matthew T. Ballew, Matthew H. Goldberg, et al., "Personal Stories Can Shift Climate Change Beliefs and Risk Perceptions: The Mediating Role of Emotion," *Communication Reports* 33, no. 3 (August 2020): 121–35.
22. David Robson, "The '3.5% Rule:' How a Small Minority Can Change the World," BBC, May 13, 2019, https://www.bbc.com/future/article/20190513-it-only-takes-35-of-people-to-change-the-world.
23. Robert O. Keohane and David G. Victor, "The Regime Complex for Climate Change," *Perspectives on Politics* 19, no. 1 (March 2011): 7.
24. Chris Castro, personal interview, August 1, 2017.
25. Michael Brune, personal interview, July 8, 2021.
26. Brune interview.
27. Heather Smith, "How to Convince the Climate Slacker to Get Serious," *Grist*, April 27, 2016, https://grist.org/climate-energy/how-to-convince-the-climate-slacker-to-get-serious.

BIBLIOGRAPHY

Abatzoglou, John T., and A. Park Williams. "Impact of Anthropogenic Climate Change on Wildfire Across Western US Forests." *Proceedings of the National Academy of Sciences* 113, no. 42 (October 10, 2016): 11,770–75. https://www.pnas.org/content/113/42/11770.

Abbey, Edward. *The Monkey Wrench Gang*. New York: Avon Books, 1975.

Adams, Mark, dir. *Raging Water*. South Carolina ETV, 2016.

Adams, Mark. *Tip of the Iceberg: My 3,000-Mile Journey Around Wild Alaska, the Last Great American Frontier*. New York: Dutton, 2018.

Adow, Mohamed. "The Climate Debt: What the West Owes the Rest." *Foreign Affairs* 99, no. 3 (May/June 2020): 60–68.

Albeck-Ripka, Livia. "Climate Change Brought a Lobster Boom. Now It Could Cause a Bust." *New York Times*, June 21, 2018. https://www.nytimes.com/2018/06/21/climate/maine-lobsters.html.

Albright, Charlotte, and Amy Olson. "Study: As Climate Changes, So Will Maple Syrup Production." *Dartmouth News*, October 2, 2019. https://news.dartmouth.edu/news/2019/10/study-climate-changes-so-will-maple-syrup-production.

Alcindor, Yamiche. "In Sweltering South, Climate Change is Workplace Hazard." *New York Times*, August 4, 2017.

Aldy, Joseph. "Money for Nothing: The Case for Eliminating US Fossil Fuel Subsidies." *Resources*, Spring/Summer 2014. https://www.resourcesmag.org/archives/money-for-nothing-the-case-for-eliminating-us-fossil-fuel-subsidies/.

Alter, Charlotte, Suyin Haynes, and Justin Worland. "Person of the Year 2019: Greta Thunberg." *Time*, December 11, 2019. https://time.com/person-of-the-year-2019-greta-thunberg/.

Anderson, Sam. "Searching for the Fountain of Youth." *New York Times*, October 24, 2014. https://www.nytimes.com/2014/10/26/magazine/my-search-for-the-fountain-of-youth.html.

Architecture 2030. "The 2030 Challenge." https://architecture2030.org/2030_challenges/2030-challenge/.

Armstrong, Anne K., Marianne E. Krasny, and Jonathon P. Schuldt. *Communicating Climate Change: A Guide for Educators*. Ithaca, NY: Comstock Publishing Associates, 2018.

Arrhenius, Svante. "On the Influence of Carbonic Acid in the Air Upon the Temperature of the Ground." *Philosophical Magazine and Journal of Science* 41, no. 5 (April 1896): 237–76.

Azevedo, Ines, Michael R. Davidson, Jesse D. Jenkins, Valerie J. Karplus, and David G. Victor. "The Paths to Net Zero: How Technology Can Save the Planet." *Foreign Affairs* 99, no. 3 (May/June 2020): 18–27.

Baker, James A. III, George Pl Shultz, and Ted Halstead. "The Strategic Case for U.S. Climate Leadership: How Americans Can Win with a Pro-Market Solution." *Foreign Affairs* 99, no. 3 (May/June 2020): 28–38.

Banerjee, Neela, Lisa Song, and David Hasemyer. "Exxon: The Road Not Taken." *Inside Climate News*, September 16, 2015. https://insideclimatenews.org/news/15092015/Exxons-own-research-confirmed-fossil-fuels-role-in-global-warming.

Barber, Michael J., and Nolan McCarty. "Causes and Consequences of Polarization." In *Solutions to Political Polarization in America*, ed. Nathaniel Persily, 15–58. New York: Cambridge University Press, 2015.

Barton, Eric. "In Sunshine State, Big Energy Blocks Solar Power." Florida Center for Investigative Reporting, April 3, 2015. https://fcir.org/2015/04/03/in-sunshine-state-big-energy-blocks-solar-power/.

Bears, Edwin C., and John C. Paige. *Historic Structure Report for Castillo de San Marcos National Monument, St. Johns County, Florida*. Denver, CO: National Park Service, U.S. Department of the Interior, September 1983. https://home.nps.gov/casa/learn/management/upload/CASA-Historic-Structure-Report-Bearss.pdf.

Bechtel, Michael M., Kenneth F. Scheve, and Elisabeth van Lieshout. "Constant Carbon Pricing Increases Support for Climate Action Compared to Ramping Up Costs Over Time." *Nature Climate Change* 10 (September 21, 2020): 1004–9. https://www.nature.com/articles/s41558-020-00914-6.

Belleville, Bill. *Losing It All to Sprawl: How Progress Ate My Cracker Landscape*. Gainesville: University Press of Florida, 2006.

Berkelmans, Ray, Glenn De'ath, Stuart Kininmonth, and William J. Skirving. "A Comparison of the 1998 and 2002 Coral Bleaching Events on the Great Barrier Reef: Spatial Correlation, Patterns, and Predictions." *Coral Reefs* 23, no. 1 (April 2004): 74–83.

Bernauer, Thomas, and Liam F. McGrath. "Simple Reframing Unlikely to Boost Public Support for Climate Policy." *Nature Climate Change* 6 (July 2016): 680–83.

Bishop, Bill. *The Big Sort: Why the Clustering of Like-Minded America Is Tearing Us Apart*. New York: Mariner Books, 2008.

Bloomberg, Michael, and Carl Pope. *Climate of Hope: How Cities, Businesses, and Citizens Can Save the Planet*. New York: St. Martin's Press, 2017.

Boehlert, Sherwood. "Can the Party of Reagan Accept the Science of Climate Change?" *Washington Post*, November 19, 2010. https://www.washingtonpost.com/wp-dyn/content/article/2010/11/18/AR2010111805451.html.

Bonica, Adam. "Data Science for the People." In *Solutions to Political Polarization in America*, ed. Nathaniel Persily, 167–77. New York: Cambridge University Press, 2015.

Bord, Robert J., Ann Fisher, and Robert E. O'Connor. "Public Perceptions of Global Warming: US and International Perspectives." *Climate Change* 11, (1998): 75–84.

Borenstein, Seth, and Emily Swanson. "AP-NORC Poll: Americans Want Local Leaders to Fight Warming." *Seattle Times*, October 2, 2017. https://www.seattletimes.com/nation-world/ap-norc-poll-americans-want-local-leaders-to-fight-warming/.

Borlaug, Brennan, Shawn Salisbury, Mindy Gerdes, and Matteo Muratori. "Levelized Cost of Charging Electric Vehicles in the United States." *Joule* 4, no. 7 (July 2020): 1470–85. https://doi.org/10.1016/j.joule.2020.05.013.

Borneman, Walter R. *Alaska: Saga of a Bold Land*. New York: HarperCollins, 2003.

Boule, Margie. "Pearl District's Namesake Was a Jewel of a Woman." *Oregonian*, April 14, 2002. https://www.oregonlive.com/portland/2002/04/pearl_districts_namesake_was_a.html.

Brinklow, Adam. "It Costs $38,000 to Create One Parking Space in San Francisco." *Curbed SF*, June 8, 2016. https://sf.curbed.com/2016/6/8/11890176/it-costs-38000-to-create-one-parking-space-in-sf.

Broecker, Wallace S. "Climatic Change: Are We on the Brink of Pronounced Global Warming?" *Science* 189, no. 4201 (August 8, 1975): 460–63.

Bromwich, Jonah Engel. "Extreme Heat Scorches Southern Arizona." *New York Times*, June 26, 2017.

Brown, E. "A Conversation on Conservation." *Washington Post*, September 8, 2016.

Brown, Lester R. *Eco-Economy: Building an Economy for the Earth*. New York: Norton, 2001.

Brown, Lester R. *World on the Edge: How to Prevent Environmental and Economic Collapse*. New York: Norton, 2011.

Brown, Taylor Kate. "Florida's Amendment 1: A Cautionary Tale for 2018?" *BBC News*, May 27, 2017. https://www.bbc.com/news/world-us-canada-39258421.

Brown, William E. *Denali: Symbol of the Alaskan Wild*. Virginia Beach, VA: Donning, 1993.

Bruce, Jeff. "Surviving the Arctic Tundra: A Look at Cold-Weather Adaptations." U.S. Department of the Interior, Bureau of Land Management Alaska. http://www.blm.gov/ak/st/en/res/education/akcold_desert/akcold desert/posterback.print.html.

Brulle, Robert J., Jason Carmichael, and J. Craig Jenkins. "Shifting Public Opinion on Climate Change: An Empirical Assessment of Factors Influencing Concern Over Climate Change in the U.S." *Climatic Change* 114, no. 2 (2012): 169–88.

Brulle, Robert J., and J. Craig Jenkins. "Fixing the Bungled US Environmental Movement." *Contexts* 7, no. 2 (2008): 14–18.

Bullard, Robert D., M. Gardezi, C. Chennault, and H. Dankbar. "Climate Change and Environmental Justice: A Conversation with Dr. Robert Bullard." *Journal of Critical Thought and Praxis* 5, no. 2 (2016). http://lib.dr.iastate.edu/jctp/vol5/iss2/3/.

Bullard, Robert D., and Beverly Wright, eds. *Race, Place, and Environmental Justice after Hurricane Katrina: Struggles to Reclaim, Rebuild, and Revitalize New Orleans and the Gulf Coast*. Boulder, CO: Westview Press, 2009.

Burke, M., S. Hsiang, S., and E. Miguel. "Global Non-Linear Effect of Temperature on Economic Production." *Nature* 527 (2015): 235–39. https://doi.org/10.1038/nature15725.

Busch, Akiko. *The Incidental Steward: Reflections on Citizen Science.* New Haven, CT: Yale University Press, 2013.

Caesar, L., G. D. McCarthy, D. J. R. Thornalley, N. Cahill, and S. Rahmstorf. "Current Atlantic Meridional Overturning Circulation Weakest in Last Millennium." *Nature Geoscience* 14 (February 25, 2021): 118–20. https://doi.org/10.1038/s41561-021-00699-z.

Caesar, L., S. Rahmstorf, A. Robinson, G. Feulner, and V. Saba. "Observed Fingerprint of a Weakening Atlantic Ocean Overturning Circulation." *Nature* 556 (April 12, 2018): 191–209.

Cagle, Alison. "Florida's Waters Are Rising, but So Are Its People." Earthjustice, January 17, 2020. https://earthjustice.org/blog/2020-january/florida-climate-justice-resiliency.

California Academy of Sciences. "American Adults Flunk Basic Science." March 13, 2009. http://www.sciencedaily.com/releases/2009/03/090312115133.htm.

California Climate & Agriculture Network. "Climate Threats to Agriculture." https://calclimateag.org/climatethreatstoag/.

California Environmental Protection Agency. "Key Events in the History of Air Quality in California." February 4, 2014. http://www.arb.ca.gov/html/brochure/history.htm.

Callendar, G. S. "The Artificial Production of Carbon Dioxide and Its Influence on Climate." *Quarterly Journal of Royal Meteorological Society* 64 (1938): 223–40.

Cann, Heather W., and Leigh Raymond. "Does Climate Denialism Still Matter? The Prevalence of Alternative Frames in Opposition to Climate Policy." *Environmental Politics* 27, no. 3 (2018): 433–54.

Cappucci, Matthew. "Tornado Alley in the Plains is an Outdated Concept. The South Is Even More Vulnerable, Research Shows." *Washington Post*, May 16, 2020. https://www.washingtonpost.com/weather/2020/05/16/tornado-alley-flawed-concept/.

Carrington, Damian. "Fossil Fuels Subsidised by $10m a Minute, Says IMF." *The Guardian*, May 18, 2015. http://www.theguardian.com/environment/2015/may/18/fossil-fuel-companies-getting-10m-a-minute-in-subsidies-says-imf.

Carson, Rachel. *Silent Spring*. New York: Houghton Mifflin, 1962.

Carter, Jimmy. "Crisis of Confidence." Speech, Washington, DC, July 15, 1979. American Rhetoric, https://www.americanrhetoric.com/speeches/jimmycartercrisisofconfidence.htm.

Caviness, Tod. "Bolting Across the State." *OrlandoSignature.com*, September 2017: 38–41.

Chakrabarty, Dipesh. "The Climate of History: Four Theses." *Critical Inquiry* 35, no. 2 (Winter 2009): 197–222.

Chandler, D. S., D. Manski, C. Donahue, and A. Alyokhin. "Biodiversity of the Schoodic Peninsula: Results of the Insect and Arachnid Bioblitzes at the Schoodic District of Acadia National Park, Maine." Maine Agricultural and Forest Experiment Station, University of Maine, Orono, 2012.

Chapman, Daniel A., Brian Lickel, and Ezra M. Markowitz. "Reassessing Emotion in Climate Change Communication." *Nature Climate Change* 7 (November 27, 2017): 850–52.

Cheatham, Amelia. "Central Florida Weather Forecast: Hot and Hotter." *Orlando Sentinel*, July 12, 2017.

Chokshi, Niraj. "Biden's Push for Electric Cars: $174 Billion, 10 Years and a Bit of Luck." *New York Times*, March 31, 2021. https://www.nytimes.com/2021/03/31/business/biden-electric-vehicles-infrastructure.html.

Chong, Dennis, and James N. Druckman. 2007. "Framing Theory." *Annual Review of Political Science* 10, no. 1 (June 15, 2007): 103–26. http://www.annualreviews.org/doi/abs/10.1146/annurev.polisci.10.072805.103054.

Christianson, Gale E. *Greenhouse: The 200-Year Story of Global Warming*. New York: Penguin, 1999.

Clarkin, Mary. "Barnes: From Big Well to City Hall." *The Hutchinson News*, November 29, 2018. https://www.hutchnews.com/news/20181126/barnes-from-big-well-to-city-hall.

Clemente, Jud. "The Great Texas Wind Power Boom." *Forbes*, October 11, 2016. https://www.forbes.com/sites/judeclemente/2016/10/11/the-great-texas-wind-power-boom/?sh=25399dc6c6aa.

Clifton, Kelly J., Sara Morrisey, and Chloe Ritter. "Business Cycles: Catering to the Bicycling Market." *TR News*, May–June 2012: 29.

Coady, David, Ian Parry, Nghia-Piotr Le, and Baoping Shang. "Global Fossil Fuel Subsidies Remain Large: An Update Based on Country-Level

Estimates." IMF Working Paper No. 19/89 (May 2, 2019). https://www.imf.org/en/Publications/WP/Issues/2019/05/02/Global-Fossil-Fuel-Subsidies-Remain-Large-An-Update-Based-on-Country-Level-Estimates-46509.

Connors, Philip. *Fire Season: Field Notes from a Wilderness Lookout*. New York: HarperCollins, 2011.

Conway, Chris. "Yellowstone's Wolves Save Its Aspens." *New York Times*, August 5, 2007. https://www.nytimes.com/2007/08/05/weekinreview/05basic.html.

Conway, Jan. "Volume of Bottled Water in the U.S. 2010–2019." *Statista*, November 26, 2020. https://www.statista.com/statistics/237832/volume-of-bottled-water-in-the-us/.

Cook, John, Dana Nuccitelli, Sarah A. Green, Mark Richardson, Bärbel Winkler, Rob Painting, Robert Way, Peter Jacobs, and Andrew Skuce. "Quantifying the Consensus on Anthropogenic Global Warming in the Scientific Literature." *Environmental Research Letters* 8, no. 2 (May 15, 2013). http://iopscience.iop.org/1748-9326/8/2/024024.

Copenhaver, John, and Joan Frye. "Solar Homeownership Panel—All About Solar." Solar United Neighbors Webinar. September 29, 2020.

Cox, Stan, and Paul Cox. "A Rising Tide: Miami Is Sinking Beneath the Sea—But Not Without a Fight." *The New Republic*, November 8, 2015. https://newrepublic.com/article/123216/miami-sinking-beneath-sea-not-without-fight.

Creswell-Myatt, Nadine. "13 Tips for Eco-Friendly Travel." *Travel Awaits*, October 14, 2020. https://www.travelawaits.com/2556711/tips-for-eco-friendly-travel/.

Cronon, William, ed. *Uncommon Ground: Toward Reinventing Nature*. New York: Norton, 1995.

Crossley, Penelope. *Renewable Energy Law: An International Assessment*. Cambridge: Cambridge University Press, 2019.

Crutzen, P. J. "The Possible Importance of COS for the Sulfate Layer of the Stratosphere." *Geophysical Research Letters* 3 (1976): 73–76.

Crutzen, Paul J. "Geology of Mankind." *Nature*, January 3, 2002: 23.

Crutzen, Paul J., and Eugene F. Stoermer. "The Anthropocene." *International Geosphere-Biosphere Programme Newsletter* 41 (May 2000): 17–18.

Dahl, Taylor. "6 Surprising Hearth Attack Triggers." *Sharecare*, September 24, 2018. https://www.sharecare.com/health/heart-attack/slideshow/6-surprising-heart-attack-triggers.

Daly, Herman. *Beyond Growth*. Boston: Beacon Press, 1996.

Darling, Seth B., and Douglas L. Sisterson. *How to Change Minds About Our Changing Climate: Let Science Do the Talking the Next Time Someone Tries to Tell You . . . and Other Arguments It's Time to End for Good*. New York: The Experiment, 2014.

Davenport, Coral. "Economies Can Still Rise as Carbon Emissions Fade." *New York Times*, April 7, 2016.

Davenport, Coral. "Talking About Climate Change with Al Gore." *Times Talks*, July 19, 2017. https://www.nytimes.com/2017/07/19/climate/al-gore-climate-change-timestalks.html.

Davis, Jack E. *The Gulf: The Making of an American Sea*. New York: Liveright, 2017.

Dawson, Ashley. *Extreme Cities: The Peril and Promise of Urban Life in the Age of Climate Change*. London: Verso, 2017.

Dean, Cornelia. *Against the Tide: The Battle for America's Beaches*. New York: Columbia University Press, 1999.

DeFries, Ruth. *What Would Nature Do? A Guide for Our Uncertain Time*. New York: Columbia University Press, 2020.

DeWeerdt, Sarah. "Why Does Climate Denial Still Receive More News Coverage than Climate Action?" *Anthropocene*, July 28, 2020. https://www.anthropocenemagazine.org/2020/07/climate-denial-still-makes-headlines/?utm_source=rss&utm_medium=rss&utm_campaign=climate-denial-still-makes-headlines.

Dewey, John. *Experience and Education*. New York: Collier, 1938.

Diamond, Jared. *Collapse: How Societies Choose to Fail or Succeed*. New York: Viking, 2005.

Dickinson, Tim. "The Climate Killers: Meet the 17 Polluters and Deniers Who Are Derailing Efforts to Curb Global Warming." *Rolling Stone*, January 6, 2010.

Dickinson, Tim. "The Koch Brothers' Dirty War on Solar Power." *Rolling Stone*, February 11, 2016. https://www.rollingstone.com/politics/politics-news/the-koch-brothers-dirty-war-on-solar-power-193325/.

Diffenbaugh, Noah S., Deepti Singh, Justin S. Mankin, Daniel E. Horton, Daniel L. Swain, Danielle Touma, Allison Charland, et al. "Quantifying the Influence of Global Warming on Unprecedented Extreme Climate Events." *Proceedings of the National Academy of Sciences* 114, no. 19 (May 9, 2017): 4881–86. https://www.pnas.org/content/114/19/4881.

Dorr, George B. *The Story of Acadia National Park: The Complete Memoir of the Man Who Made It All Possible*. Bar Harbor, ME: Acadia, 1997.

Doucleff, Michaeleen. "Is There a Ticking Time Bomb Under the Arctic? *NPR Morning Edition*, January 24, 2018. https://www.npr.org/sections/goatsandsoda/2018/01/24/575220206/is-there-a-ticking-time-bomb-under-the-arctic.

Doyle, Rice. "Fla. Gov. Bans the Terms 'Climate Change,' 'Global Warming.'" *USA Today*, March 9, 2015.

Drouin, Roger. "How Conservative Texas Took the Lead in U.S. Wind Power." *Yale Environment 360*, April 9, 2015. https://e360.yale.edu/features/how_conservative_texas_took_the_lead_in_us_wind_power.

Duany, Andrés, Elizabeth Plater-Zyberk, and Jeff Speck. *Suburban Nation: The Rise of Sprawl and the Decline of the American Dream*. New York: North Point Press, 2000.

Dunlap, Riley E., Aaron M. McCright, and Jerrod H. Yarosh. "The Political Divide on Climate Change: Partisan Polarization Widens in the U.S." *Environment* 58, no. 5 (September 2, 2016): 4–23.

Earthwatch. *Climate Change: Sea to Trees at Acadia National Park*. Boston: Earthwatch Institute, 2018.

Eckhouse, Brian. "World Added More Solar, Wind Than Anything Else Last Year." *Bloomberg*, September 1, 2020. https://www.bloomberg.com/news/articles/2020-09-01/the-world-added-more-solar-wind-than-anything-else-last-year.

Egan, Patrick J., and Megan Mullin. "Climate Change: U.S. Public Opinion." *Annual Review of Political Science* 20 (May 2017): 209–227. https://doi.org/10.1146/annurev-polisci-051215-022857.

Ellsmoor, James. "United States Spends Ten Times More on Fossil Fuel Subsidies than Education." *Forbes*, June 15, 2019. https://www.forbes.com/sites/jamesellsmoor/2019/06/15/united-states-spend-ten-times-more-on-fossil-fuel-subsidies-than-education/?sh=36fa834b4473.

Emanuel, Kerry. *Divine Wind: The History and Science of Hurricanes*. New York: Oxford University Press, 2005.

Emanuel, Kerry. *What We Know About Climate Change*. Cambridge, MA: MIT Press, 2012.

Englander, John. *High Tide on Main Street: Rising Sea Level and the Coming Coastal Crisis*. Boca Raton, FL: Science Bookshelf, 2012.

Epstein, Paul R., and Gary M. Tabor. "Climate Change is Really Bugging Our Forests." *Washington Post*, September 7, 2003. https://www.washingtonpost.com/archive/opinions/2003/09/07/climate-change-is-really-bugging-our-forests/2ea6f965-1f73-4805-b9a1-e3f38cf0640a/.

Ezenwa, Vanessa O., David J. Civitello, Brandon T. Barton, Daniel J. Becker, Maris Brenn-White, Aimee T. Classen, Sharon L. Deem, et al. "Infectious Diseases, Livestock, and Climate: A Vicious Cycle?" *Trends in Ecology and Evolution* 35, no. 11 (November 1, 2020): 959–62. https://doi.org/10.1016/j.tree.2020.08.012.

Ferguson, Gary. *Land on Fire: The New Reality of Wildfire in the West*. Portland, OR: Timber Press, 2017.

Fiaschi, Simone, and Shimon Wdowinski. "Local Land Subsidence in Miami Beach (FL) and Norfolk (VA) and Its Contribution to Flooding Hazard in Coastal Communities Along the U.S. Atlantic Coast." *Ocean & Coastal Management* 187 (April 1, 2020). https://doi.org/10.1016/j.ocecoaman.2019.105078.

Figueres, Christiana, and Tom Rivett-Carnac. *The Future We Choose: The Stubborn Optimist's Guide to the Climate Crisis*. New York: Vintage, 2021.

Fischer, Douglas. "'Dark Money' Funds Climate Change Denial Effort." *Scientific American*, December 23, 2013. http://www.scientificamerican.com/article/dark-money-funds-climate-change-denial-effort/?print=true.

Fister, Barbara. "More on Alaska." *Inside Higher Ed*, July 3, 2019. https://www.insidehighered.com/blogs/library-babel-fish/more-alaska.

Flannery, Tim. *Atmosphere of Hope: Searching for Solutions to the Climate Crisis*. New York: Atlantic Monthly Press, 2015.

Flannery, Tim. *The Weathermakers: How Man is Changing the Climate and What It Means for Life on Earth*. New York: Atlantic Monthly Press, 2005.

Flannigan, M. D., and C. E. van Wagner. "Climate Change and Wildfire in Canada." *Canadian Journal of Forest Research* 21 (1991): 66–72.

Flavelle, Christopher. "How California Became Ground Zero for Climate Disasters." *New York Times*, September 20, 2020. https://www.nytimes.com/2020/09/20/climate/california-climate-change-fires.html.

Florida Department of Environmental Protection. "Florida's Coral Reefs." August 5, 2020. https://floridadep.gov/rcp/rcp/content/floridas-coral-reefs.

Florida, Richard. *The Rise of the Creative Class, Revisited*. New York: Basic Books, 2011.

Foer, Jonathan Safran. *We Are the Weather: Saving the Planet Begins at Breakfast*. New York: Farrar, Straus, and Giroux, 2019.

Fothergill, Alastair, and Jonathan Hughes, dirs. *David Attenborough: A Life on Our Planet*. 2020.

Fountain, Henry. "Warmer Alaska May Mean More Carbon Emissions." *New York Times*, May 9, 2017.

Fourier, Joseph. "Remarques générales sur les températures du globe terrestre et des espaces planétaires." *Annales de Chemie et de Physique* 27 (1824): 136–67.

Fox, Thomas J. *Green Town U.S.A.: The Handbook for America's Sustainable Future*. Hobart, NY: Hatherleigh, 2013.

Fraga, Robert. *The Greening of Oz: Sustainable Architecture in the Wake of a Tornado*. Shelbyville, KY: Wasteland Press, 2012.

Friedman, Lisa. "Fixing a Major Piece of the Climate Puzzle." *New York Times*, July 14, 2017.

Friedman, Lisa. "Governor of California Plans Summit on Climate." *New York Times*, July 7, 2017.

Friedman, Lisa. "Greenhouse Gases Increase, but Scientists at NOAA Say, 'It's Complicated.'" *New York Times*, July 14, 2017.

Friedman, Thomas. *Hot, Flat, and Crowded: Why We Need a Green Revolution—and How It Can Renew America*. New York: Farrar, Straus, and Giroux, 2008.

Friedman, Thomas L. "Paris Climate Accord is a Big, Big Deal." *New York Times*, December 16, 2015. http://www.nytimes.com/2015/12/16/opinion/paris-climate-accord-is-a-big-big-deal.html.

Friedrichs, Jörg. *The Future is Not What It Used to Be: Climate Change and Energy Security*. Cambridge, MA: MIT Press, 2013.

Fuller, Thomas, and Christopher Flavelle. "A Climate Reckoning in Fire-Stricken California." *New York Times*, October 27, 2020. https://www.nytimes.com/2020/09/10/us/climate-change-california-wildfires.html.

Funk, Cary, and Brian Kennedy. "How Americans See Climate Change and the Environment in 7 Charts." *Pew Research Center*, April 21, 2020. https://www.pewresearch.org/fact-tank/2020/04/21/how-americans-see-climate-change-and-the-environment-in-7-charts/.

Galbraith, Kate, and Asher Price. *The Great Texas Wind Rush: How George Bush, Ann Richards, and a Bunch of Tinkerers Helped the Oil and Gas State Win the Race to Wind Power.* Austin: University of Texas Press, 2013.

Gallegos, Jenna. "Rising Temperatures Could Bump You from Your Flight. Thanks, Climate Change." *Washington Post*, July 3, 2017.

Gammon, Crystal. "Changing Climate Increases West Nile Threat in U.S." *East Bay Times*, March 20, 2009, https://www.eastbaytimes.com/2009/03/20/changing-climate-increases-west-nile-threat-in-u-s/.

Garcia, Sierra. "Here Are the Top Ways the World Could Take on Climate Change in 2020." *Grist*, March 19, 2020. https://grist.org/climate/here-are-the-top-ways-the-world-could-take-on-climate-change-in-2020.

Gardner, Sheldon. "Advocacy Group Part of St. Augustine's Sea Level Rise Efforts." *St. Augustine Record*, March 26, 2017. http://staugustine.com/news/local-news/2017-03-26/advocacy-group-part-st-augustine-s-sea-level-rise-efforts.

Gattuso, Jean-Pierre, and Lina Hansson. "Ocean Acidification: Background and History." In *Ocean Acidification*, ed. Jean-Pierre Gattuso and Lina Hansson, 1–20. Oxford: Oxford University Press, 2011.

Gelbspan, Ross. *Boiling Point: How Politicians, Big Oil and Coal, Journalists, and Activists Have Fueled the Climate Crisis—and What We Can Do to Avert Disaster.* New York: Basic Books, 2004.

Gertner, Jon. "Should the United States Save Tangier Island from Oblivion." *New York Times Magazine*, July 6, 2016. https://www.nytimes.com/2016/07/10/magazine/should-the-united-states-save-tangier-island-from-oblivion.html.

Gibbens, Sarah. "How Warm Oceans Supercharge Deadly Hurricanes." *National Geographic*, September 4, 2019. https://www.nationalgeographic.com/environment/article/how-warm-water-fuels-a-hurricane.

Giddens, Anthony. *The Politics of Climate Change.* Cambridge: Polity Press, 2009.

Gifford, Robert. "The Dragons of Inaction: Psychological Barriers that Limit Climate Change Mitigation and Adaptation." *American Psychologist* 66, no. 4 (May–June 2011): 290–302.

Gillis, Justin, and Jugal K. Patel. "Antarctica Sheds Huge Iceberg that Hints at Future Calamity." *New York Times*, July 13, 2017.

Giordono, L., H. Boudet, and A. Gard-Murray. "Local Adaptation Policy Responses to Extreme Weather Events." *Policy Sciences* 53 (August 18, 2020): 609–36. https://doi.org/10.1007/s11077-020-09401-3.

Gladwell, Malcolm. *The Tipping Point: How Little Things Can Make a Big Difference*. Boston: Little, Brown & Company, 2000.

Glynn, Peter W. "Global Warming and Widespread Coral Mortality: Evidence of First Coral Reef Extinctions." In *Saving a Million Species: Extinction Risk form Climate Change*, ed. Lee Hannah, 103–20. Washington, DC: Island Press, 2012.

Godoy, Maria. "In U.S. Cities, The Health Effects of Past Housing Discrimination Are Plain to See." NPR, November 19, 2020. https://www.npr.org/sections/health-shots/2020/11/19/911909187/in-u-s-cities-the-health-effects-of-past-housing-discrimination-are-plain-to-see.

Goldberg, Rebecca F., and Laura N. Vandenberg. "Distract, Delay, Disrupt: Examples of Manufactured Doubt from Five Industries." *Reviews on Environmental Health* 34, no. 4 (2019): 349–63. https://doi.org/10.1515/reveh-2019-0004.

Goldenberg, Suzanne. "US East Coast Cities Face Frequent Flooding Due to Climate Change." *The Guardian*, October 8, 2014. https://www.theguardian.com/environment/2014/oct/08/us-east-coast-cities-face-frequent-flooding-due-to-climate-change.

Goodell, Jeff. "Goodbye, Miami." *Rolling Stone*, July 4, 2013.

Goodell, Jeff. *The Water Will Come: Rising Seas, Sinking Cities, and the Remaking of the Civilized World*. New York: Little, Brown & Company, 2017.

Goodrich, David. *A Hole in the Wind: A Climate Scientist's Bicycle Journey across the United States*. New York: Pegasus Books, 2017.

Gore, Al. "The Case for Optimism on Climate Change." TED Vancouver, February 16, 2016. http://www.ted.com/talks/al_gore_the_case_for_optimism_on_climate_change.

Gough, C., and S. Shackley. "The Respectable Politics of Climate Change: The Epistemic Communities and NGOs." *International Affairs* 77, no. 2 (2001): 329–45.

Gowen, Annie. "The Town That Built Back Green." *Washington Post*, October 23, 2020. https://www.washingtonpost.com/climate-solutions/2020/10/22/greensburg-kansas-wind-power-carbon-emissions/.

Greene, C. W., L. L. Gregory, G. H. Mittelhauser, S. C. Rooney, and J. E. Weber. "Vascular Flora of the Acadia National Park Region, Maine." *Rhodora* 107 (2005):117–85.

Grescoe, Taras. *Strap Hanger: Saving Our Cities and Ourselves from the Automobile.* New York: Times Books/Henry Holt and Company, 2012.

Grunwald, Michael. "Progmation: Why Hurricane Katrina was a Man-Made Disaster." *New Republic*, August 14–21, 2006: 32–37.

Guggenheim, Davis, dir. *An Inconvenient Truth: A Global Warning.* Paramount Pictures, 2006.

Gunter, Mike, Jr. "Confront Climate Change or Count on Ever Crueler Hurricanes and Septembers." *USA Today*, October 4, 2017, https://www.usatoday.com/story/opinion/2017/10/04/confront-climate-change-or-count-on-ever-crueler-hurricanes-septembers-mike-gunter-column/724671001/.

Gunter, Mike, Jr. *Tales of an Ecotourist: What Travel to Wild Places Can Teach Us about Climate Change.* Albany: SUNY Press, 2018.

Gustafson, Abel, Matthew T. Ballew, Matthew H. Goldberg, Matthew J. Cutler, Seth A. Rosenthal, and Anthony Leiserowitz. "Personal Stories Can Shift Climate Change Beliefs and Risk Perceptions: The Mediating Role of Emotion." *Communication Reports* 33, no. 3 (August 2020): 121–35.

Gutschke, Laura. "Rain Not Going Away This Week for Abilene." *Abilene Reporter-News*, May 19, 2021.

Guzman, Andrew. *Overheated: The Human Cost of Climate Change.* Oxford: Oxford University Press, 2013.

Hacker, Jacob S., and Paul Pierson. "Confronting Asymmetric Polarization." In *Solutions to Political Polarization in America*, ed. Nathaniel Persily, 59–72. New York: Cambridge University Press, 2015.

Hall, Shannon. "Exxon Knew About Climate Change Almost 40 Years Ago." *Scientific American*, October 26, 2015. http://www.scientificamerican.com/article/exxon-knew-about-climate-change-almost-40-years-ago/.

Hancock, Elaina. "UConn Research: More Carbon in the Ocean Can Lead to Smaller Fish." *UConn Today*, August 4, 2020. https://today.uconn.edu/2020/08/uconn-research-carbon-ocean-can-lead-smaller-fish/#.

Hansen, James. *Storms of My Grandchildren: The Truth About the Coming Climate Catastrophe and Our Last Chance to Save Humanity.* New York: Bloomsbury, 2009.

Hansen, James. "Why I Must Speak Out About Climate Change." TED, February 2012.

Hansen, J., I. Fung, A. Lacis, D. Rind, S. Lebedeff, R. Ruedy, G. Russell, et al. "Global Climate Changes as Forecast by Goddard Institute for Space Studies Three-Dimensional Model." *Journal of Geophysical Research* 93, no. D8 (August 20, 1988): 9341–64.

Hansen, James, Makiko Sato, Paul Hearty, Reto Ruedy, Maxwell Kelley, Valerie Masson Delmotte, Gary Russell, et al. "Ice Melt, Sea Level Rise and Superstorms: Evidence from Paleoclimate Data, Climate Modeling, and Modern Observations That 2° C Global Warming Could Be Dangerous." *Atmospheric Chemistry and Physics* 16 (March 22, 2016): 3761–812. https://acp.copernicus.org/articles/16/3761/2016/acp-16-3761-2016.pdf.

Hardin, Garrett. "The Tragedy of the Commons." *Science* 162 (1968): 1243–48.

Hargrove, Brantley. "47 Feet High and Rising." *Nashville Scene*, April 28–May 4, 2011.

Harris, Alex, and Joey Flechas. "Miami Could Eliminate Climate Change Czar as COVID Forces Budget Chops Across City." *Miami Herald*, July 23, 2020. https://www.miamiherald.com/news/local/environment/article244433242.html.

Hart, P. S., and E. C. Nisbet. "Boomerang Effects in Science Communication: How Motivated Reasoning and Identity Cues Amplify Opinion Polarization About Climate Mitigation Policies." *Communication Research* 39, no. 6 (2012): 701–23. https://doi.org/10.1177/0093650211416646.

Hawken, Paul. "Beck Environmental Lecture." Fall for the Book, George Mason University, October 10, 2018.

Hawken, Paul. *Blessed Unrest: How the Largest Movement in the World Came Into Being and Why No One Saw it Coming*. New York: Viking, 2007.

Hawken, Paul. *Drawdown*. New York: Penguin Books, 2017.

Hawken, Paul. *The Ecology of Commerce: A Declaration of Sustainability*. New York: HarperBusiness, 1993.

Hayes, Christopher. "The New Abolitionism: Averting Planetary Disaster Will Mean Forcing Fossil Fuel Companies to Give Up at Least $10 Trillion in Wealth." *The Nation*, May 12, 2014.

Hayhoe, Katharine. *Saving Us: A Climate Scientist's Case for Hope and Healing in a Divided World*. New York: One Signal Publishers/Atria Books, 2021.

Henderson, Rebecca. "The Unlikely Environmentalists: How the Private Sector Can Combat Climate Change." *Foreign Affairs* 99, no. 3 (May/June 2020): 47–52.

Henry, Tom. "Climate Change Called Certain and Most Predictions Are Bad." *Toledo Blade*, October 13, 2008.

Heeren, Fredric. "Rebuilding Greensburg Green." Smithsonianmag.com, February 27, 2009. https://www.smithsonianmag.com/science-nature/rebuilding-greensburg-green-55848425/.

Hineline, Mark L. *Ground Truth: A Guide to Tracking Climate Change at Home*. Chicago: University of Chicago Press, 2018.

Hoegh-Guldberg, O. "Coral Reefs, Climate Change, and Mass Extinction." In *Saving a Million Species: Extinction Risk from Climate Change*, ed. Lee Hannah, 261–84. Washington, DC: Island Press, 2012.

Hoffman, Jeremy. "Throwing Shade in RVA." http://jeremyscotthoffman.com/throwing-shade.

Hoffman, J. S., V. Shandas, and N. Pendleton. "The Effects of Historical Housing Policies on Resident Exposure to Intra-Urban Heat: A Study of 108 US Urban Areas." *Climate* 8, no. 1 (2020): 12. https://doi.org/10.3390/cli8010012.

Hofstadter. Richard. *Anti-Intellectualism in American* Life. New York: Knopf, 1963.

Hollander, Zaz. "Boat Explosion and Fire Leaves Part of Whittier Dock Unsafe and Hundreds of Pounds of Fish in Limbo." *Anchorage Daily News*, July 11, 2019. https://www.adn.com/alaska-news/2019/07/11/boat-explosion-and-fire-leaves-part-of-whittier-dock-unsafe-and-hundreds-of-pounds-of-fish-in-limbo/.

Hollander, Zaz. "Coast Guard Suspends Search for Man Missing After Propane Blast at Whittier Dock." *Anchorage Daily News*, July 8, 2019. https://www.adn.com/alaska-news/2019/07/08/two-missing-after-explosion-aboard-fishing-vessel-at-whittier-dock/#.

Hond, Paul. "The Ice Detectives." *Columbia Magazine*, Fall 2017, 12–21.

Horn, Miriam. *Rancher, Farmer, Fisherman: Conservation Heroes of the American Heartland*. New York: Norton, 2016.

Horne, Jed. *Breach of Faith: Hurricane Katrina and the Near Death of a Great American City*. New York: Random House, 2008.

Hoxie, G. L. "How Fire Helps Forest." *Sunset* 34 (1910): 145–51.

Hu, Winnie. "In New York, Rush Hour Comes to the Bike Lane." *New York Times*, July 31, 2017.

Hughes, Sara. *Repowering Cities: Governing Climate Change Mitigation in New York City, Los Angeles, and Toronto*. Ithaca, NY: Cornell University Press, 2019.

Hulme, Mike. *Why We Disagree About Climate Change: Understanding Controversy, Inaction and Opportunity*. Cambridge: Cambridge University Press, 2009.

Hutchins, Cory. "A Laurel to WLTX Meteorologist Jim Gandy." *Columbia Journalism Review*, March 7, 2013. https://archives.cjr.org/united_states_project/a_laurel_to_wltx_meteorologist_jim_gandy_in_south_carolina.php.

Ingraffea, Anthony. "Gangplank to a Warm Future." *New York Times*, July 28, 2013. https://www.nytimes.com/2013/07/29/opinion/gangplank-to-a-warm-future.html.

Inhofe, James M. "The Science of Climate Change." Floor Statement by Sen. Inhofe (R-OK), U.S. Senate. July 28, 2003.

Intergovernmental Panel on Climate Change (IPCC). "Climate Change 2014 Synthesis Report Summary for Policymakers." https://www.ipcc.ch/report/ar5/syr/.

Intergovernmental Panel on Climate Change (IPCC). "Fifth Assessment Report (AR5)." http://www.ipcc.ch/report/ar5/.

Jaccard, Mark. *The Citizen's Guide to Climate Success: Overcoming Myths That Hinder Progress*. Cambridge: Cambridge University Press, 2020.

Jackson, Kenneth T. *Crabgrass Frontier: The Suburbanization of the United States*. New York: Oxford University Press, 1985.

Jacobs, Jane. *The Death and Life of Great American Cities*. New York: Random House, 1961.

Jacques, Peter J., Riley E. Dunlap, and Mark Freeman. "The Organization of Denial: Conservative Think Tanks and Environmental Skepticism." *Environmental Politics* 17, no. 3 (June 2008): 349–85.

Jahren, Hope. *The Story of More: How We Got to Climate Change and Where to Go from Here*. New York: Vintage Books, 2020.

Jenkins, Mark. "New Program Helps Low-Income D.C. Homeowners Convert to Solar Energy." *Washington Post*, May 27, 2015. https://www.washingtonpost.com/local/new-program-helps-low-income-dc

-homeowners-convert-to-solar-energy/2015/05/26/adoce104-03b2-11e5-bc72-f3e16bf50bb6_story.html.

Jevons, William Stanley. *The Coal Question: An Enquiry Concerning the Progress of the Nation, and the Probable Exhaustion of Our Coal-mines*. London: Macmillan, 1865.

Johnson, Julie, Omar Shaikh Rashad, and Matthias Gafni. "Dixie Fire Explodes Near Paradise, Site of the Devastating 2018 Camp Fire." *San Francisco Chronicle*, July 14, 2021. https://www.sfchronicle.com/california-wildfires/article/Dixie-Fire-near-Paradise-explodes-to-1-200-acres-16314692.php.

Johnson, Nathanael. "Low on Water, California Farmers Turn to Solar Farming." *Grist*, August 6, 2019. https://grist.org/article/california-farmer-solar-panel-water-renewable-nature-conservancy.

Johnstone, Bob. *Switching to Solar: What We Can Learn from Germany's Success in Harnessing Clean Energy*. Amherst, NY: Prometheus Books, 2011.

Joint Center for Housing Studies of Harvard University. "The State of the Nation's Housing 2020." https://www.jchs.harvard.edu/state-nations-housing-2020.

Kahan, Dan. "Fixing the Communications Failure." *Nature* 463, no. 21 (2010): 296–97. http://dx.doi.org/10.1038/463296a.

Kahn, Matthew E. *Green Cities: Urban Growth and the Environment*. Washington, DC: Brookings Institution Press, 2006.

Kalmus, Peter. *Being the Change: Live Well and Spark a Climate Revolution*. Gabriola Island, BC: New Society Publishers, 2017.

Kamarck, Elaine C. "Solutions to Polarization." In *Solutions to Political Polarization in America*, ed. Nathaniel Persily, 96–103. New York: Cambridge University Press, 2015.

Kamp, David. "Can Miami Beach Survive Global Warming?" *Vanity Fair*, November 10, 2015. https://www.vanityfair.com/news/2015/11/miami-beach-global-warming.

Karelas, Andreas. *Climate Courage: How Tackling Climate Change Can Build Community, Transform the Economy, and Bridge the Political Divide in America*. Boston: Beacon Press, 2020.

Kaufman, Rachel. "These Are the Most Bikeable Cities in America." NextCity.org, May 16, 2018. https://nextcity.org/daily/entry/these-are-the-most-bikeable-cities-in-america.

Kay, Jane Holtz. *Asphalt Nation: How the Automobile Took Over America and How We Can Take It Back*. Berkley: University of California Press, 1998.

Keane, Phoebe. "How the Oil Industry Made Us Doubt Climate Change." BBC News, September 19, 2020.

Kelley, David. "Symposium: Sub-Tropical and Tropical Coastal Resilience: Social, Economic, and Physical Adaptations in South Florida and the Caribbean." University of Miami, Coral Gables, FL, April 7, 2021. https://resilience.miami.edu/index.html.

Kelley, Tyler. *Holding Back the River: The Struggle Against Nature on America's Waterways*. New York: Avid Reader Press, 2021.

Kendrick, Daisy. *The Climate Is Changing: Why Aren't We?* London: Piatkus, 2020.

Keohane, Robert O., and David G. Victor. "The Regime Complex for Climate Change." *Perspectives on Politics* 19, no. 1 (March 2011): 7–23.

Key, V. O. *Public Opinion and American Democracy*. New York: Knopf, 1961.

Key, V. O. *The Responsible Electorate*. Cambridge, MA: Harvard University Press, 1966.

Khanna, Pankaj, André W. Droxler, Jeffrey A. Nittrouer, John W. Tunnell Jr., and Thomas C. Shirley. "Coralgal Reef Morphology Records Punctuated Sea-Level Rise During the Last Deglaciation." *Nature Communications* 8, no. 1046 (October 19, 2017). https://www.nature.com/articles/s41467-017-00966-x.

Klein, Naomi. *This Changes Everything: Capitalism vs the Climate*. New York: Simon & Schuster, 2014.

Knudson, Tom. "Sierra Warming, Later Snow, Earlier Melt." *Sacramento Bee*, December 26, 2008.

Kolbert, Elizabeth. *Field Notes from a Catastrophe: Man, Nature, and Climate Change*. New York: Bloomsbury, 2006.

Kopestinsky, Alex. "Electric Car Statistics in the US and Abroad." *Policy Advice*, April 6, 2021. https://policyadvice.net/insurance/insights/electric-car-statistics/.

Krakauer, Jon. *Into the Wild*. New York: Villard, 1996.

Krisher, Tom. "US Report: Gas Mileage Down with Emissions Up." *Orlando Sentinel*, January 7, 2021.

Krugman, Paul. "Building a Green Economy." *New York Times Magazine*, April 7, 2010. http://www.nytimes.com/2010/04/11/magazine/11Economy-t.html.

Krupp, Fred, and Miriam Horn. *Earth: The Sequel*. New York: Norton, 2008.

Kupers, Roland. *A Climate Policy Revolution: What the Science of Complexity Reveals About Saving Our Planet*. Cambridge, MA: Harvard University Press, 2020.

Kurz, W. A. "Mountain Pine Beetle and Forest Carbon Feedback." *Nature* 452 (April 24, 2008): 987–90.

Kusnetz, Nicholas. "South Miami Approves Solar Roof Rules, Inspired by a Teenager." *Inside Climate News*, July 18, 2017. https://insideclimatenews.org/news/18072017/south-miami-florida-solar-roof-rules-sea-level-rise-teenager-activism/.

Langdon, Philip. *Within Walking Distance: Creating Livable Communities for All*. Washington, DC: Island Press, 2017.

Lanza, Michael. *Before They're Gone: A Family's Year-Long Quest to Explore America's Most Endangered National Parks*. Boston: Beacon Press, 2012.

Law, Tara. "About 2.5 Million Acres in Alaska Have Burned. The State's Wildfire Seasons Are Getting Worse, Experts Say." *Time*, August 20, 2019. https://time.com/5657188/alaska-fires-long-climate-change/.

Law, Kara Lavender, Natalie Starr, Theodore R. Siegler, Jenna R. Jambeck, Nicholas J. Mallos, and George H. Leonard. "The United States' Contribution of Plastic Waste to Land and Ocean." *Science Advances* 6, no. 44 (October 30, 2020). https://advances.sciencemag.org/content/6/44/eabd0288.

Lawler, Joshua J., D. Scott Rinnan, Julia L. Michalak, John C. Withey, Christopher R. Randels, and Hugh P. Possingham. "Planning for Climate Change Through Additions to a National Protected Area Network: Implications for Cost and Configuration." *Philosophical Transactions of the Royal Society B*, January 27, 2020. https://royalsocietypublishing.org/doi/10.1098/rstb.2019.0117.

Leefeldt, Ed, and Amy Danise. "FEMA'S Upcoming Changes Could Cause Flood Insurance to Soar at the Shore." *Forbes*, March 18, 2021. https://www.forbes.com/advisor/homeowners-insurance/new-fema-flood-insurance-rates/.

Leonard, Annie. *The Story of Stuff: How Our Obsession with Stuff Is Trashing the Planet, Our Communities, and Our Health—and a Vision for Change.* New York: Free Press, 2010.

Leopold, Aldo. *A Sand County Almanac: And Sketches Here and There.* New York: Oxford University Press, 1949.

Levi, Michael. *The Power Surge: Energy, Opportunity, and the Battle for America's Future.* New York: Oxford University Press, 2013.

Levitsky, Steven, and Daniel Ziblatt. *How Democracies Die.* New York: Crown, 2018.

Li, Lin, and Pinaki Chakraborty. "Slower Decay of Landfalling Hurricanes in a Warming World." *Nature* 587, (2020): 230–34. https://doi.org/10.1038/s41586-020-2867-7.

Lijphart, Arend. "Polarization and Democratization." In *Solutions to Political Polarization in America*, ed. Nathaniel Persily, 73–82. New York: Cambridge University Press, 2015.

Lipset, Seymour Martin. *Political Man: The Social Bases of Politics.* Garden City, NY: Doubleday, 1960.

Lomborg, Bjorn. *Cool It: The Skeptical Environmentalist's Guide to Global Warming.* New York: Knopf, 2007.

Lord, Nancy. *Green Alaska: Dreams from the Far Coast.* Washington, DC: Counterpoint, 1999.

Lorenz, Taylor. "Are Gender Reveals Cursed?" *New York Times*, September 10, 2020. https://www.nytimes.com/2020/09/10/style/gender-reveal-parties-cursed.html.

Loria, Kevin. "Miami Is Racing Against Time to Keep Up with Sea-Level Rise." *Business Insider*, April 12, 2018. https://www.businessinsider.com/miami-floods-sea-level-rise-solutions-2018-4.

Louisiana Office of Community Development. "Resettlement of Isle de Jean Charles: Background & Overview." June 9, 2020. https://isledejeancharles.la.gov/sites/default/files/public/IDJC-Background-and-Overview-1-28-21.pdf.

Louv, Richard. *Last Child in the Woods: Saving Our Children from Nature-Deficit Disorder.* Chapel Hill, NC: Algonquin Books, 2005.

Lovelock, J. E. *GAIA: A New Look at Life on Earth.* Oxford: Oxford University Press, 1979.

Lovelock, James. *The Revenge of GAIA: Earth's Climate Crisis & the Fate of Humanity.* New York: Basic Books, 2006.

Lovins, Amory B. "Energy Strategy: The Road Not Taken?" *Foreign Affairs* 55, no. 1 (October 1976): 65–96.

Lustgarten, Abrahm. "How Climate Migration Will Reshape America." *New York Times Magazine*, September 20, 2020. https://www.nytimes.com/interactive/2020/09/15/magazine/climate-crisis-migration-america.html.

Lutsey, Nicholas, and Daniel Sperling. "America's Bottom-Up Climate Change Mitigation Policy." *Energy Policy* 36 (February 2008): 673–85.

Lydersen, Kari. "Risk of Disease Rises with Water Temperatures." *Washington Post*, October 20, 2008.

Lynas, Mark. *High Tide: The Truth About Our Climate Crisis*. New York: Picador, 2004.

Ma, Michelle. "Rethinking Land Conservation to Protect Species that Will Need to Move with Climate Change." *UW News*, January 28, 2020. https://www.washington.edu/news/2020/01/28/rethinking-land-conservation-to-protect-species-that-will-need-to-move-with-climate-change/.

Maclean, Norman. *Young Men and Fire*. Chicago: University of Chicago Press, 1992.

Macnab, Deirdre. "12-Year-Old Starts Solar Solution: Neighborhood Rooftop Solar Co-ops Driving Expansion State by State." *The Solutions Journal* 10, no. 2 (May 2019). https://www.thesolutionsjournal.com/article/12-year-old-starts-solar-solution-neighborhood-rooftop-solar-co-ops-driving-expansion-state-state.

Magill, Bobby. "South Miami Just Made a Huge Rooftop Solar Decision." *Climate Central*, July 20, 2017. https://www.climatecentral.org/news/florida-california-solar-mandate-21631.

Maibach, E., J. Witte, and K. Wilson. "Climate-Gate Undermined Belief in Global Warming Among Many TV Meteorologists." *Bulletin of the American Meteorological Association* 92 (2011): 31–37.

Maldonado, Charles. "Rollin' Down the River." *City Paper*, May 22–24, 2010, 12–13.

Maniates, Michael. "Going Green? Easy Doesn't Do It." *Washington Post*, November 22, 2007.

Manjoo, Farhad. "How Y2K Offers a Lesson for Fighting Climate Change." *New York Times*, July 19, 2017: B1, 9. https://www.nytimes.com/2017/07/19/technology/y2k-lesson-climate-change.html.

Manjoo, Farhad. "I've Seen a Future Without Cars, and It's Amazing." *New York Times*, July 9, 2020.

Mann, Michael E. "Climate Scientists Feel Your Pain, Dr. Fauci." *Newsweek*, August 11, 2020. https://www.newsweek.com/climate-scientists-feel-your-pain-dr-fauci-opinion-1524293.

Mann, Michael E. *The Hockey Stick and the Climate Wars: Dispatches from the Front Lines*. New York: Columbia University Press, 2012.

Mann, Michael E., Raymond S. Bradley, and Malcolm K. Hughes. "Global-Scale Temperature Patterns and Climate Forcing Over the Past Six Centuries." *Nature* 392 (April 23, 1998): 779–87.

Mann, Michael E., and Peter H. Gleick. "Climate Change and California Drought in the 21st Century." *Proceedings of the National Academy of Sciences* 112 (2015): 3858–59.

Mann, Michael E., Stefan Rahmstorf, Kai Kornhuber, Byron A. Steinman, Sonya K. Miller, and Dim Coumou. "Influence of Anthropogenic Climate Change on Planetary Wave Resonance and Extreme Weather Events." *Scientific Reports* 7, no. 45242 (2017). https://doi.org/10.1038/srep45242.

Mann, Michael E., and Tom Toles. *The Madhouse Effect: How Climate Change Denial Is Threatening Our Planet, Destroying Our Politics, and Driving Us Crazy*. New York: Columbia University Press, 2016.

Mapes, Jeff. *Pedaling Revolution: How Cyclists Are Changing American Cities*. Corvallis: Oregon State University Press, 2009.

Marchese, David. "Greta Thunberg Hears Your Excuses. She Is Not Impressed." *New York Times Magazine*, October 30, 2020. https://www.nytimes.com/interactive/2020/11/02/magazine/greta-thunberg-interview.html.

Marlon, Jennifer. "7 Ways You're Already Paying for Climate Change." *Barron's*, September 13, 2020. https://www.barrons.com/articles/7-ways-youre-already-paying-for-climate-change-51599995430.

Marlon, Jennifer, and Abigail Cheskis. "Wildfires and Climate Are Related—Are Americans Connecting the Dots?" Yale Program on Climate Change Communication Blog, December 11, 2017. https://climatecommunication.yale.edu/news-events/connecting-wildfires-with-climate/.

Marlon, Jennifer, Peter Howe, Matto Mildenberger, Anthony Leiserowitz, and Xinran Wang. "Yale Climate Opinion Maps 2020." Yale Program on Climate Change Communication. September 2, 2020. https://climatecommunication.yale.edu/visualizations-data/ycom-us/.

Marsden, William. *Fools Rule: Inside the Failed Politics of Climate Change.* Toronto: Knopf Canada, 2011.

Marshall, George. *Don't Even Think About It: Why Our Brains Are Wired to Ignore Climate Change.* New York: Bloomsbury, 2014.

Marsooli, Reza, Ning Lin, Kerry Emmanuel, and Kairui Feng. "Climate Change Exacerbates Hurricane Flood Hazards Along US Atlantic and Gulf Coasts in Spatially Varying Patterns." *Nature Communications* 10, no. 3,785 (2019). https://doi.org/10.1038/s41467-019-11755-z.

Massachusetts et al. v. EPA et al. No. 05-1120. Supreme Court of the United States. 415 F.3d 50, April 2, 2007. Legal Information Institute, Cornell University Law School. https://www.law.cornell.edu/supct/html/05-1120.ZS.html.

Mayer, Jane. *Dark Money: The Hidden History of the Billionaires Behind the Rise of the Radical Right.* New York: Doubleday, 2016.

McBeath, Jerry, Matthew Bermna, Jonathan Rosenberg, and Mary F. Ehrlander. *The Political Economy of Oil in Alaska: Multinationals vs. the State.* Boulder, CO: Lynne Rienner, 2008.

McCarty, Nolan. "Reducing Polarization by Making Parties Stronger." In *Solutions to Political Polarization in America*, ed. Nathaniel Persily, 136–45. New York: Cambridge University Press, 2015.

McCright, Aaron M., and Riley E. Dunlap. "The Politicization of Climate Change and Polarization in the American Public's Views of Global Warming, 2001–2010." *The Sociological Quarterly* 52, no. 2 (March 2011): 155–94.

McKibben, Bill. "The Climate Crisis." *The New Yorker*, November 4, 2020.

McKibben, Bill. *Eaarth: Making a Life on a Tough New Planet.* New York: Times Books, 2010.

McKibben, Bill. *The End of Nature.* New York: Random House, 1989.

McKibben, Bill. "Global Warming's Terrifying New Math: Three Simple Numbers That Add Up to Global Catastrophe—and That Make Clear Who the Real Enemy Is." *Rolling Stone*, July 19, 2012.

McKibben, Bill. "The Nature of Crisis." *The New Yorker*, March 26, 2020. https://www.newyorker.com/news/annals-of-a-warming-planet/the-nature-of-crisis-coronavirus-climate-change.

McKibben, Bill. "What Will It Take to Cool the Planet?" *The New Yorker*, May 21, 2020. https://www.newyorker.com/news/annals-of-a-warming-planet/what-will-it-take-to-cool-the-planet .

McKnight, Michael. "Paradise Lost and Found." *Sports Illustrated*, November 4, 2019, 64–71.

McLeod, Kathy Baughman. "Building a Resilient Planet: How to Adapt to Climate Change from the Bottom-Up." *Foreign Affairs* 99, no. 3 (May/June 2020): 54–59.

McMullen, Jane, dir. *Frontline: Fire in Paradise*. PBS, 2019.

Mikhitarian, Sarah. "Homes With Solar Panels Sell for 4.1% More." Zillow, April 16, 2019. https://www.zillow.com/research/solar-panels-house-sell-more-23798/

Miller, Alesa. "Greensburg: Is the City Still Green and Sustainable?" *Kiowa County Signal*, January 16, 2019. https://www.kiowacountysignal.com/news/20190116/greensburg-is-city-still-green-and-sustainable.

Miller-Rushing, Abe. "Managing a Changing Acadia National Park." Acadia and Schoodic Education and Research Center, National Park Service, U.S. Department of Interior, June 25, 2018.

Miller-Rushing, A. J., T. L. Lloyd-Evans, R. B. Primack, and P. Satzinger. "Bird Migration Times, Climate Change, and Changing Population Sizes." *Global Change Biology* 14, (2008): 1959–72.

Milman, Oliver. "Atlantic City and Miami Beach: Two Takes on Tackling the Rising Waters." *The Guardian*, March 20, 2017. https://www.theguardian.com/us-news/2017/mar/20/atlantic-city-miami-beach-sea-level-rise.

Mooney, Chris. "The Hockey Stick: The Most Controversial Chart in Science, Explained." *The Atlantic*, May 10, 2013. http://www.theatlantic.com/technology/archive/2013/05/the-hockey-stick-the-most-controversial-chart-in-science-explained/275753/.

Mooney, Chris. "What We Can Say about the Louisiana Floods and Climate Change." *Washington Post*, August 15, 2016. https://www.washingtonpost.com/news/energy-environment/wp/2016/08/15/what-we-can-say-about-the-louisiana-floods-and-climate-change/.

Moore, Kathleen Dean. *Great Tide Rising: Towards Clarity and Moral Courage in a Time of Planetary Change*. Berkeley, CA: Counterpoint, 2016.

Morris, Frank. "Kansas Town's Green Dreams Could Save Its Future." *All Things Considered*, December 27, 2007. https://www.npr.org/templates/story/story.php?storyId=17643060.

Morrison, Jim. "Climate Change Turns the Tide on Waterfront Living." *Washington Post*, April 13, 2020. https://www.washingtonpost.com/magazine/2020/04/13/after-decades-waterfront-living-climate-change-is-forcing-communities-plan-their-retreat-coasts/?arc404=true.

Morton, Thomas A., Anna Rabinovich, Dan Marshall, and Pamela Bretschneider. "The Future That May (or May Not) Come: How Framing Changes Responses to Uncertainty in Climate Change Communications." *Global Environmental Change* 21, no. 1 (February 2011): 103–9.

Muir, John. *Travels in Alaska*. New York: Houghton Mifflin, 1915.

Mumford, Lewis. *The Culture of Cities*. New York: Harcourt, Brace and Company, 1938.

Muro, Mark, and Devashree Saha. "Rooftop Solar: Net Metering Is a Net Benefit." Brookings Institute, May 23, 2016. https://www.brookings.edu/research/rooftop-solar-net-metering-is-a-net-benefit/.

Murphy, Pat. *Plan C: Community Survival Strategies for Peak Oil*. Gabriola Island, BC: New Society, 2008.

Murphy, Zoeann, and Chris Mooney. "Gone in a Generation: Across America, Climate Change Is Already Disrupting Lives." *Washington Post*, January 29, 2019. https://www.washingtonpost.com/graphics/2019/national/gone-in-a-generation/.

Murray, Christopher S., and Hannes Baumann. "Are Long-Term Growth Responses to Elevated pCO_2 Sex-Specific in Fish?" *PLoS One* 15, no. 7 (July 17, 2020). https://doi.org/10.1371/journal.pone.0235817.

Myers, Andrew. "Stanford Study Finds that Wind Energy Output Increases When People Need Heat the Most." *Stanford/News*, March 17, 2021. https://news.stanford.edu/2021/03/17/coldest-times-wind-energy-production-heats/.

NASA. "Understanding Sea Level—Thermal Expansion." https://sealevel.nasa.gov/understanding-sea-level/global-sea-level/thermal-expansion.

National Oceanic and Atmospheric Administration. "What Is a Perigean Spring Tide?" https://oceanservice.noaa.gov/facts/perigean-spring-tide.html.

National Conference of State Legislatures. "State Renewable Portfolio Standards and Goals." January 4, 2020. https://www.ncsl.org/research/energy/renewable-portfolio-standards.aspx#nmi.

Nelson, Robert K., LaDale Winling, Richard Marciano, Nathan Connolly, et al. "Mapping Inequality." In *American Panorama*, ed. Robert K. Nelson and Edward L. Ayers. https://dsl.richmond.edu/panorama/redlining/.

Netburn, Deborah. "Sierra Nevada Snowpack on Track to Shrink Up to 79% by the End of the Century, New Study Finds." *Los Angeles Times*, December 16, 2018.

Newell, P. "Climate for Change? Civil Society and the Politics of Global Warming." In *Global Civil Society 2005/6*, ed. M. Glasius, M. Kaldor, and H. Anheir, 90–199. Thousand Oaks, CA: Sage, 2005.

Newman, Peter, Timothy Beatley, and Heather Boyer. *Resilient Cities: Responding to Peak Oil and Climate Change*. Washington, DC: Island Press, 2009.

Nicholson, Simon. "Solar Radiation Management." Wilson Center, September 30, 2020. https://www.wilsoncenter.org/article/solar-radiation-management#footnote3.

Nisbet, Matthew C. "Communicating Climate Change: Why Frames Matter for Public Engagement." *Environment: Science and Policy for Sustainable Development* 51, no. 2 (2009): 12–23.

Nistor, M. M., and I. M. Petcu. "Quantitative Analysis of Glaciers Changes from Passage Canal Based on GIS and Satellite Images, South Alaska." *Applied Ecology and Environmental Research* 13, no. 2 (2015): 535–49.

NOAA National Centers for Environmental Information. "Billion-Dollar Weather and Climate Disasters: Mapping." July 9, 2021. https://www.ncdc.noaa.gov/billions/mapping.

Nordhaus, Ted, and M. Shellenberger. "The End of Magical Climate Thinking." *Foreign Policy*, January 13, 2010. https://foreignpolicy.com/2010/01/13/the-end-of-magical-climate-thinking/.

Nordhaus, William. "The Climate Club: How to Fix a Failing Global Effort." *Foreign Affairs* 99, no. 3 (May/June 2020): 10–17.

"Norfolk Vision 2100." City of Norfolk. Adopted November 22, 2016.

Nossiter, Adam. "An Empty Place Where a Kansas Town Once Stood." *New York Times*, May 7, 2007. https://www.nytimes.com/2007/05/07/us/07tornado.html.

"The Numbers." *Portland Business Journal*, May 18, 2018.

Ockwell, David, Lorraine Whitmarsh, and Saffron O'Neill. "Reorienting Climate Change Communication for Effective Mitigation: Forcing People to Be Green or Fostering Grass-Roots Engagement?" *Science Communication* 30, no. 3 (January 7, 2009): 305–27. https://doi.org/10.1177/1075547008328969.

Oliver, John. "Climate Change Debate." *Last Week Tonight with John Oliver*, HBO, May 11, 2014.

Oliver, John. "Green New Deal." *Last Week Tonight with John Oliver*, HBO, May 12, 2019.

Oltmanns, M., F. Straneo, and M. Tedesco. "Increased Greenland Melt Triggered by Large-Scale, Year-Round Cyclonic Moisture Intrusions." *The Cryosphere* 13 (2019): 815–25. https://doi.org/10.5194/tc-13-815-2019.

Oppenheimer, Michael. "As the World Burns: Climate Change's Dangerous Next Phase." *Foreign Affairs* 99, no. 6 (November/December 2020): 34–40.

Oregon Institute of Science and Medicine. "The Global Warming Petition Project." http://www.petitionproject.org/.

Oreskes, Naomi. "The Scientific Consensus on Climate Change." *Science* 306, no. 5702 (December 3, 2004): 1686. https://science.sciencemag.org/content/306/5702/1686.

Oreskes, Naomi, and Erik M. Conway. *Merchants of Doubt: How a Handful of Scientists Obscured the Truth on Issues from Tobacco Smoke to Global Warming*. New York: Bloomsbury Press, 2010.

Orlowski, Jeff, dir. *Chasing Coral*. Netflix, 2017.

Orr, David W. *Down to the Wire: Confronting Climate Collapse*. Oxford: Oxford University Press, 2009.

Orr, David W. *Earth in Mind: On Education, Environment, and the Human Prospect*. Washington, DC: Island Press, 1994.

Osaka, Shannon. "Good News: Americans Can Freak Out About Coronavirus and Climate Change at the Same Time." *Grist*, May 20, 2020. https://grist.org/climate/good-news-americans-can-freak-out-about-coronavirus-and-climate-change-at-the-same-time/.

Osaka, Shannon. "How the US Power Grid Could Go On a 90% Carbon-Free Diet." *Grist*, June 10, 2020. https://grist.org/energy/how-the-us-electricity-grid-could-go-on-a-90-carbon-free-diet.

Osaka, Shannon. "'The Planet Is Broken,' UN Chief Says." *Grist*, December 3, 2020. https://grist.org/climate/the-planet-is-broken-u-n-leader-says-in-state-of-the-climate-report.

Osaka, Shannon. "Will the West's Giant Fires Spark a Climate Awakening?" *Grist*, September 14, 2020. https://grist.org/climate/will-the-wests-california-wildfire-season-change-minds-on-climate-change.

Ostrom, Elinor. *Governing the Commons: The Evolution of Institutions for Collective Action*. New York: Cambridge University Press, 1990.

Owen, David. *Green Metropolis: Why Living Smaller, Living Closer, and Driving Less Are the Keys to Sustainability*. New York: Riverhead Books, 2009.

Owens, Paul. "Toll-Road Plans Fall Short on Wildlife Protection, Urban Sprawl." *Orlando Sentinel*, November 15, 2020.

Pacalo, Stephen, and Robert Sucohow. "Stabilization Wedges: Solving the Climate Problem for the Next 50 Years with Current Technologies." *Science* 305, no. 5686 (August 13, 2004): 968–72. https://science.sciencemag.org/content/305/5686/968.

Packer, George. "The Four Americas." *The Atlantic*, July/August 2021. https://www.theatlantic.com/magazine/archive/2021/07/george-packer-four-americas/619012/.

Palm, Risa, Toby Bolsen, and Justin T. Kingsland. "'Don't Tell Me What to Do': Resistance to Climate Change Messages Suggesting Behavior Changes." *Weather, Climate, and Society* 12, no. 4 (2020): 827–35. https://doi.org/10.1175/WCAS-D-19-0141.1.

"Paradise Nature-Based Fire Resilience Project Final Report." Paradise Recreation and Park District, June 2020.

Parker, Laura. "How Megafires Are Remaking American Forests." *National Geographic*, August 9, 2015. https://www.nationalgeographic.com/science/article/150809-wildfires-forest-fires-climate-change-science.

Parker, Laura. "Hurricane Matthew's Destructive Storm Surges Hint at New Normal." *National Geographic*, October 8, 2016. https://www.nationalgeographic.com/science/article/hurricane-matthew-storm-surges-predict-sea-level-rise-btf.

Parkes, Graham. *How to Think About the Climate Crisis: A Philosophical Guide to Saner Ways of Living*. New York: Bloomsbury, 2021.

Parmesan, Camille, and Gary Yohe. "A Globally Coherent Fingerprint of Climate Change Impacts Across Natural Systems." *Nature* 421 (January 2, 2003): 37–42.

Patel, Prachi. "Rooftop Solar Panels Could Provide Nearly Half of Our Power." *Conservation Magazine*, April 14, 2016. https://www.conservationmagazine.org/2016/04/rooftop-solar-panels-provide-nearly-half-power/.

Pathak, Tapan B., Mahesh L. Maskey, Jeffery A. Dahlberg, Faith Kearns, Khaled M. Ball, and Daniele Zaccaria. "Climate Change Trends and Impacts on California Agriculture: A Detailed Review." *Agronomy* 8, no. 3 (February 26, 2018). https://doi.org/10.3390/agronomy8030025.

Patrick, Katie. "Urban Heat Islands: The Secret Killer You've Never Heard Of, with Jeremy Hoffman, Ph.D." *How to Save the World Podcast*, April 30, 2018.

Pearce, Warren, Brian Brown, Brigitte Nerlich, and Nelya Koteyko. "Communicating Climate Change: Conduits, Content, and Consensus." *Wiley Interdisciplinary Reviews: Climate Change* 6, no. 6 (2015): 613–26.

Pearson, Richard. *Driven to Extinction: The Impact of Climate Change on Biodiversity*. New York: Sterling, 2011.

Pedersen, Joe Mario. "Central Florida Hit by EF-2 Tornado with 115 mph Winds, NWS Reports." *Orlando Sentinel*, August 20, 2020.

Penn, Ivan. "The Sunshine Statement." *New York Times*, August 31, 2018.

Perkins, John H. *Changing Energy: The Transition to a Sustainable Future*. Berkeley: University of California Press, 2017.

Persily, Nathaniel. "Introduction." In *Solutions to Political Polarization in America*, ed. Nathaniel Persily, 3–14. New York: Cambridge University Press, 2015.

Peters, Ian Marius, Christopher Brabec, Tonio Buonassisi, Jens Hauch, and Andre M. Nobre. "The Impact of COVID-19-Related Measures on the Solar Resource in Areas with High Levels of Air Pollution." *Joule* 4, no. 8 (August 19, 2020): 1681–87. https://www.sciencedirect.com/science/article/abs/pii/S2542435120302725.

Phillimore, John, and Aidan Davison. "A Precautionary Tale: Y2K and the Politics of Foresight." *Futures* 34, no. 2 (2002): 147–57.

Pielke, Roger, Jr. *The Climate Fix: What Scientists and Politicians Won't Tell You About Global Warming*. New York: Basic Books, 2010.

Piligian, Craig, dir. *Greensburg: A Story of Community Rebuilding*. Discovery Communications, 2009.

Pilkey, Orrin H., and Rob Young. *The Rising Sea*. Washington, DC: Island Press, 2009.

Pittman, Craig. *Oh, Florida! How America's Weirdest State Influences the Rest of the Country*. New York: St. Martin's Press, 2016.

Plass, Gilbert N. "The Carbon Dioxide Theory of Climatic Change." *Tellus* 8, no. 2 (1956): 140–54.

Plumer, Brad. "Assessing the Economic Bite from Rising Temperatures." *New York Times*, June 30, 2017.

Plumer, Brad, and Nadja Popovich. "How Decades of Racist Housing Policy Left Neighborhoods Sweltering." *New York Times*, August 24, 2020. https://www.nytimes.com/interactive/2020/08/24/climate/racism-redlining-cities-global-warming.html.

Podesta, John, and Todd Stern. "A Foreign Policy for the Climate: How American Leadership Can Avert Catastrophe." *Foreign Affairs* 99, no. 3 (May/June 2020): 39–46.

"Politico 50: Our Guide to the Thinkers, Doers and Visionaries Transforming American Politics in 2016." *Politico Magazine*. https://www.politico.com/magazine/politico50/2016/philip-stoddard-harold-wanless/.

Poole, Leslie Kemp. *Biscayne National Park: The History of a Unique Park on the "Edge."* Washington, DC: National Park Service, 2022.

Pooley, Eric. *The Climate War: True Believers, Power Brokers, and the Fight to Save the Earth*. New York: Hyperion, 2010.

Pope Francis. "Encyclical Letter Laudato Si' of the Holy Father Francis on Care for Our Common Home." Vatican: The Holy See. Rome, June 18, 2015. http://w2.vatican.va/content/francesco/en/encyclicals/documents/papa-francesco_20150524_enciclica-laudato-si.pdf.

Popovich, Nadja, and Brad Plumer. "How Does Your State Make Electricity." *New York Times*, October 28, 2020. https://www.nytimes.com/interactive/2020/10/28/climate/how-electricity-generation-changed-in-your-state-election.html.

Primack, R. B., and A. J. Miller-Rushing. "Uncovering, Collecting, and Analyzing Records to Investigate the Ecological Impacts of Climate Change: A Template from Thoreau's Concord." *BioScience* 62 (2012): 170–81.

Prior, Markus, and Natalie Jomini Stroud. "Using Mobilization, Media, and Motivation to Curb Political Polarization." In *Solutions to Political Polarization in America*, ed. Nathaniel Persily, 78–194. New York: Cambridge University Press, 2015.

Proctor, Caitlin R., Juneseok Lee, David Yu, Amisha D. Shah, and Andrew J. Whelton. "Wildfire Caused Widespread Drinking Water Distribution

Network Contamination." *Water Science* 2, no. 4 (July/August 2020). https://awwa.onlinelibrary.wiley.com/doi/full/10.1002/aws2.1183.

Public Financing of Construction Projects. 2020 Florida Legislature SB 178, Section 161.551, Florida Statutes, effective July 1, 2021. https://flsenate.gov/Session/Bill/2020/178/BillText/er/HTML.

Pulle, Matt, and Liz Garrigan. "Up the Creek." *Nashville Scene* 29, no. 23 (July 8–14, 2010): 9–15.

Putnam, Robert D. *Bowling Alone: The Collapse and Revival of American Community*. New York: Simon & Schuster, 2000.

Pyper, Julia. "Florida Voters Defeat Utility-Backed Solar Amendment." *Green Tech Media*, November 9, 2016. https://www.greentechmedia.com/articles/read/florida-voters-defeat-utility-backed-solar-amendment.

Rabe, Barry. *Can We Price Carbon?* Cambridge, MA: MIT Press, 2018.

Rabe, Barry G. *Statehouse and Greenhouse: The Emerging Politics of American Climate Change Policy*. Washington, DC: Brookings Institution Press, 2004.

Rachel Carson Council. "Colleges and Other Institutions Can Change the Food System." December 8, 2020. https://rachelcarsoncouncil.org/colleges-institutions-can-change-food-system.

Radeloff, Volker C., David P. Helmers, H. Anu Kramer, Miranda H. Mockrin, Patricia M. Alexandre, Avi Bar-Massada, Van Butsic, et al. "Rapid Growth of the US Wildland-Urban Interface Raises Wildfire Risk." *Proceedings of the National Academy of Sciences of the United States of America* 115, no. 13 (March 27, 2018): 3314–19. https://doi.org/10.1073/pnas.1718850115.

Ramirez, Rachel. "Extreme Heat Is Worse in Redlined Neighborhoods." *Grist*, August 4, 2020. https://grist.org/justice/extreme-heat-redlining-portland.

Ranck, Jessica. "Abilene Residents Unhappy with Street Drainage System, City Says There's Not Much They Can Do About It." *Big Country Homepage*, April 28, 2021. https://www.bigcountryhomepage.com/news/main-news/abilene-residents-unhappy-with-drainage-system-city-says-theres-not-much-they-can-do-about-it/.

Regional Greenhouse Gas Initiative, Inc. (RGGI). http://www.rggi.org/.

Rempel, Austin. "Replanting Paradise." *American Forests*, Fall 2020. https://www.americanforests.org/magazine/article/replanting-paradise/.

Revelle, Roger, and Hans E. Suess. "Carbon Dioxide Exchange Between Atmosphere and Ocean and the Question of an Increase of Atmospheric CO_2 During the Past Decades." *Tellus* 9 (1957): 18–27.

Revkin, Andrew. "Climate Expert Says NASA Tried to Silence Him." *New York Times*, January 29, 2006.

Revkin, Andrew C. "Paths to a 'Good' Anthropocene." Keynote Address at Association for Environmental Studies and Sciences, Pace University, New York, NY, June 11, 2014.

Rhoades, Alan M., Andrew D. Jones, and Paul A Ulrich. "The Changing Character of the California Sierra Nevada as a Natural Reservoir." *Geophysical Research Letters* 45, no. 23 (December 16, 2018): 13008–19.

Richardson, Jeff. "Chena Hot Springs Resort Owner Bernie Karl Named UAF's Business Leader of the Year." *Fairbanks Daily News-Miner*, March 28, 2010. http://www.newsminer.com/news/local_news/chena-hot-springs-resort-owner-bernie-karl-named-uaf-s/article_db69bfaa-1361-5249-9663-2d701324fa8f.html.

Ritchie, Hannah. "Cars, Planes, Trains: Where Do CO_2 Emissions from Transport Come From?" *Our World in Data*, October 6, 2020. https://ourworldindata.org/co2-emissions-from-transport.

Rivers, Brendan. "Deluged by Floods, America's 'Oldest City' Struggles to save Landmarks from Climate Crisis." *The Guardian*, October 28, 2020. https://www.theguardian.com/us-news/2020/oct/28/st-augustine-florida-floods-climate-crisis.

Robbins, Jim. "Can Geothermal Power Play a Key Role in the Energy Transition?" *Yale Environment 360*, December 22, 2020. https://e360.yale.edu/features/can-geothermal-power-play-a-key-role-in-the-energy-transition.

Roberts, David. "Florida's Outrageously Deceptive Solar Ballot Initiative, Explained. Amendment 1 Is a Utility Scam." *Vox*, November 8, 2016. https://www.vox.com/science-and-health/2016/11/4/13485164/florida-amendment-1-explained .

Roberts, David. "Is it Worth Trying to "Reframe" Climate Change? Probably Not." *Vox*, February 27, 2017. https://www.vox.com/2016/3/15/11232024/reframe-climate-change .

Robock, A. "20 Reasons Why Geoengineering May Be a Bad Idea." *Bulletin of the Atomic Scientists* 64 (2008): 14–18, 59.

Robson, David. "The '3.5% Rule': How a Small Minority Can Change the World." BBC, May 13, 2019. https://www.bbc.com/future/article/20190513-it-only-takes-35-of-people-to-change-the-world.

Rocky Mountain Research Station. "Frequently Asked Questions about the Mountain Pine Beetle Epidemic." U.S. Forest Service. https://www.fs.usda.gov/rmrs/frequently-asked-questions-about-mountain-pine-beetle-epidemic.

Rollins, Chris, Jeffrey T. Freymueller, and Jeanne M. Sauber. "Stress Promotion of the 1958 M_w~7.8 Fairweather Fault Earthquake and Others in Southeast Alaska by Glacial Isostatic Adjustment and Inter-earthquake Stress Transfer." *JGR Solid Earth*, December 11, 2020. https://agupubs.onlinelibrary.wiley.com/doi/full/10.1029/2020JB020411.

Rossi, Jason. "15 American Cities with the Most Homes in Danger of Flooding." Showbiz Cheat Sheet, December 11, 2018. https://www.cheatsheet.com/culture/american-cities-homes-danger-flooding.html/.

Rousseau, Jean-Jacques. *Emile, or On Education*. Translated by Allan Bloom. New York: Basic Books, 1979.

Ruckelshaus, William D., Lee M. Thomas, William K. Reilly, and Christine Todd Whitman. "A Republican Case for Climate Action." *New York Times*, August 1, 2013. https://www.nytimes.com/2013/08/02/opinion/a-republican-case-for-climate-action.html.

Rush, Elizabeth. *Rising: Dispatches from the New American Shore*. Minneapolis: Milkweed, 2019.

Sadik-Khan, Janette, and Seth Solomonow. *Street Fight: Handbook for an Urban Revolution*. New York: Viking, 2016.

Samenow, Jason. "Death Valley Soars to 130 Degrees, Potentially Earth's Highest Temperature Since at Least 1931." *Washington Post*, August 16, 2020. https://www.washingtonpost.com/weather/2020/08/16/death-valley-heat-record/.

Samenow, Jason. "Japan's Cherry Blossoms Signal Warmest Climate in More Than 1,000 Years." *Washington Post*, April 4, 2017. https://www.washingtonpost.com/news/capital-weather-gang/wp/2017/04/04/japans-cherry-blossoms-signal-warmest-climate-in-over-1000-years/.

Samuel, Molly. "How the Dream of America's 'Nuclear Renaissance' Fizzled." NPR: Weekend Edition Sunday, August 6, 2017. http://www.npr.org/2017/08/06/541582729/how-the-dream-of-americas-nuclear-renaissance-failed-to-materialize.

Sanford, Todd, Regina Wang, and Alyson Kenward. "The Age of Alaskan Wildfires." *Climate Central*, 2015. http://assets.climatecentral.org/pdfs/AgeofAlaskanWildfires.pdf.

Sanger, David E., and Jane Perlez. "Giving China a Void to Fill." *New York Times*, June 2, 2017.

Schellenberg, Michael, and Ted Nordhaus. "The Death of Environmentalism: Global Warming Politics in a Post-Environmental World." Essay delivered at annual meeting of Environmental Grantmakers Association, 2004.

Schendler, Auden, and Randy Udall. "LEED Is Broken; Let's Fix It." DailyReporter.com, October 2005. https://dailyreporter.com/wp-content/blogs.dir/1/files/2010/08/leedisbroken.pdf.

Schneider, Stephen H. *Science as a Contact Sport: Inside the Battle to Save Earth's Climate*. Washington, DC: National Geographic, 2009.

Schouten, Cory. "'Climate Gentrification' Could Add Value to Elevation in Real Estate." *MoneyWatch*, December 28, 2017. https://www.cbsnews.com/news/climate-gentrification-home-values-rising-sea-level/.

Schumacher, E. F. *Small Is Beautiful: Economics as if People Mattered*. London: Blond & Briggs, Ltd., 1973.

Schwartz, John, and Veronica Penney. "In the West, Lightning Grows as a Cause of Damaging Fires." *New York Times*, October 23, 2020.

Scranton, Roy. "Learning How to Die in the Anthropocene." *New York Times*, November 10, 2013.

Scranton, Roy. "Raising My Child in a Doomed World." *New York Times*, July 16, 2018.

Sellers, Frances Stead. "Charlotte Bulldozes Against Flooding." *Orlando Sentinel*, December 1, 2019.

Serafin, Tatiana. "Dumbest Business Idea of the Year." *Forbes*, July 5, 2004. https://www.forbes.com/celebrities2004/064a.html.

Shabecoff, Philip. "Global Warming Has Begun, Expert Tells Senate." *New York Times*, June 24, 1988. https://www.nytimes.com/1988/06/24/us/global-warming-has-begun-expert-tells-senate.html.

Shephard, Marshall. "Are We Experiencing a New Normal with Extreme Weather?" Climate Correction, University of Central Florida, Orlando, October 3, 2019.

Shoup, Donald. *The High Cost of Free Parking*. Chicago: University of Chicago Press, 2008.

Sibilla, Nick. "Connecticut Should Be Tesla Turf." *New York Times*, July 7, 2017.

Smil, Vaclav. "Moore's Curse and the Great Energy Delusion." *The American*, November 19, 2008.

Smith, Sean K. *You Can Save the Earth*. Long Island City, NY: Hatherleigh Press, 2008.

Southeast Florida Regional Climate Change Compact Sea Level Rise Work Group (Compact). February 2020. Document prepared for the Southeast Florida Regional Climate Change Compact Climate Leadership Committee. https://southeastfloridaclimatecompact.org/wp-content/uploads/2020/04/Sea-Level-Rise-Projection-Guidance-Report_FINAL_02212020.pdf.

Spear, Kevin. "Orlando No.1 Again for Pedestrian Deaths." *Orlando Sentinel*. January 23, 2019. https://www.orlandosentinel.com/news/transportation/os-ne-orlando-deadliest-pedestrians-worsening-20190122-story.html.

"Species on the Move: American Lobster." *Inside Climate News*. https://insideclimatenews.org/species/invertebrates/lobster.

Speck, Jeff. *Walkable City: How Downtown Can Save America, One Step at a Time*. New York: Farrar, Straus, and Giroux, 2012.

Spence, Alexa, Wouter Poortinga, and Nick Pidgeon. "The Psychological Distance of Climate Change." *Risk Analysis* 32, no. 6 (June 1, 2012): 957–72.

Spotts, Pete. "Record-Breaking Floods Force Engineers to Blow up Mississippi River Levee." *Christian Science Monitor*, May 2, 2011. https://www.csmonitor.com/USA/2011/0502/Record-breaking-floods-force-engineers-to-blow-up-Mississippi-River-levee.

"St. Augustine's Lost Seawall." *Historic Preservation* 17, no. 4 (July–August 1965): 139.

St. John, Paige, Joseph Serna, and Rong-Gon Lin II. "Must Reads: Here's How Paradise Ignored Warnings and Became a Deathtrap." *Los Angeles Times*, December 30, 2018. https://www.latimes.com/local/california/la-me-camp-fire-deathtrap-20181230-story.html.

Stein, Kate. "Temperatures in Florida Are Rising. For Vulnerable Patients, The Heat Can Be Life-Threatening." Florida Center for Investigative Reporting, November 13, 2018. https://www.wlrn.org/news/2018-11-13/temperatures-in-florida-are-rising-for-vulnerable-patients-the-heat-can-be-life-threatening.

Stephenson, R. Bruce. *Portland's Good Life: Sustainability and Hope in an American City*. New York: Lexington Books, 2021.

Stephenson, Bruce. "Portland Defines the New Decade in Planning." *Planning Magazine*, January 1, 2021. https://www.planning.org/planning/2021/winter/intersections-viewpoint/.

Stern, Nicholas. *The Economics of Climate Change: The Stern Review*. Cambridge: Cambridge University Press and HUM Treasury, 2006.

Steves, Rick. "The Value of Travel." TEDxRainer, November 12, 2011. http://tedxtalks.ted.com/video/TEDxRainier-Rick-Steves-The-Val.

Stokenes, Per Espen. *What We Think About When We Try Not to Think About Global Warming: Toward a New Psychology of Climate Action*. White River Junction, VT: Chelsea Green, 2015.

Stone, Maddie. "Want to Know What Climate Change Feels Like? Ask an Alaskan." *Grist*, October 8, 2019. https://grist.org/article/want-to-know-what-climate-change-feels-like-ask-an-alaskan.

Struzik, Edward. *Firestorm: How Wildfire Will Shape Our Future*. Washington, DC: Island Press, 2017.

"Study: Global Warming Could Boost Crop Pests." *Chicago Tribune*, December 16, 2008.

Sueur, Jerome, Bernie Krause, and Almo Farina. "Climate Change Is Breaking Earth's Beat." *Trends in Ecology & Evolution* 24, no. 11 (November 2019): 971–73.

Svolik, Milan W. "Polarization Versus Democracy." *Journal of Democracy* 30, no. 3 (July 2019): 20–32.

Symon, Evan. "Los Angeles City Council Votes to Start Ban on Single Use Plastic Water Bottles." *California Globe*, November 7, 2019. https://californiaglobe.com/section-2/los-angeles-city-council-votes-to-start-ban-on-single-use-plastic-water-bottles/.

Symons, Jonathan. *Ecomodernism: Technology, Politics and the Climate Crisis*. Cambridge: Polity Press, 2019.

Tabari, Hossein. "Climate Change Impact on Flood and Extreme Precipitation Increases with Water Availability." *Scientific Reports* 10, no. 13768 (2020). https://doi.org/10.1038.

Tamman, Maurice. "The Great Lobster Rush." *Reuters*, October 30, 2018. https://www.reuters.com/investigates/special-report/ocean-shock-lobster/.

Tedesco, Marco, and Alberto Flores d'Arcais. *The Hidden Life of Ice: Dispatches from a Disappearing World*. New York: The Experiment, 2020.

Teirstein, Zoya. "Florida Republicans Are Ready to Stop Rising Seas—Just Not Climate Change." *Grist*, March 9, 2021. https://grist.org/politics/florida-republicans-are-ready-to-stop-rising-seas-just-not-climate-change.

Teproff, Carli. "New Homes Will Now Require Solar Panels in South Miami, a First in Florida." *Miami Herald*, July 18, 2017. https://www.miamiherald.com/news/local/community/miami-dade/south-miami/article162307863.html.

Testimony of Michael Grimm, Before the Committee on Science, Space and Technology Subcommittee on Investigations and Oversight Subcommittee on Environment, 116th Congress. February 27, 2020.

Thill, David. "Maine Company Looks to Tidal Power as Renewable Energy's Next Generation." Energy News Network, September 23, 2020. https://energynews.us/2020/09/23/northeast/maine-company-looks-to-tidal-power-as-renewable-energys-next-generation/.

Thoman, R., and J. E. Walsh. *Alaska's Changing Environment: Documenting Alaska's Physical and Biological Changes Through Observations*. International Arctic Research Center, University of Alaska Fairbanks, 2019.

Thomas, J. L., C. M. Polashenski, A. J. Soja, L. Marelle, K. A. Casey, H. D. Choi, J.-C. Raut, et al. "Quantifying Black Carbon Deposition Over the Greenland Ice Sheet from Forest Fires in Canada." *Geophysical Research Letters* 44, no. 15 (August 16, 2017): 7965–74. https://agupubs.onlinelibrary.wiley.com/doi/full/10.1002/2017GL073701.

Thomasson, Scott. "Vote Solar & Business Groups Commend Governor Rick Scott for Signing onto Solar." *Vote Solar FL*, June 16, 2017. https://votesolar.org/usa/florida/updates/rick-scott-signs-solar-bill/.

Thoreau, Henry David. "Walking." *The Atlantic Monthly* 9, no. 56 (June 1862): 657–74.

Tidwell, Mike. *Bayou Farewell: The Rich Life and Tragic Death of Louisiana's Cajun Coast*. New York: Vintage, 2003.

Tomain, Joseph P. *Clean Power Politics: The Democratization of Energy*. Cambridge: Cambridge University Press, 2017.

"Top 10 Biggest Wind Farms." *Power Technology.com*, January 11, 2021. https://www.power-technology.com/features/feature-biggest-wind-farms-in-the-world-texas/.

Transport for London. *Town Centre Study*. September 2011.

Twain, Mark. *The Innocents Abroad*. Hartford, CT: American Publishing Company, 1869.

Tyson, Alec, and Brian Kennedy. "Two-Thirds of Americans Think Government Should Do More on Climate." Pew Research Center, June 23, 2020. https://www.pewresearch.org/science/2020/06/23/two-thirds-of-americans-think-government-should-do-more-on-climate/.

UNFCCC. "Paris Agreement." December 12, 2015. http://unfccc.int/files/meetings/paris_nov_2015/application/pdf/paris_agreement_english_.pdf.

Union of Concerned Scientists. "Sea Level Rise and Tidal Flooding in Norfolk, Virginia." March 30, 2016. https://ucsusa.org/resources/sea-level-rise-and-tidal-flooding-norfolk-virginia.

United Nations Economic Commission for Europe. "Sustainable Energy: Methane Management." https://unece.org/challenge.

United Nations Framework Convention on Climate Change. "Kyoto Protocol." http://unfccc.int/kyoto_protocol/items/2830.php.

Upin, Catherine, dir. *Frontline: Climate of Doubt*. PBS, 2012.

Urban Land Institute. "Charlotte-Mecklenburg Floodplain Buyout Program." https://developingresilience.uli.org/case/charlotte-mecklenburg-floodplain-buyout-program/.

U.S. Bureau of Labor Statistics. "Fastest Growing Occupations." *Occupational Outlook Handbook*, April 9, 2021. https://www.bls.gov/ooh/fastest-growing.htm.

U.S. Energy Information Administration. "Today in Energy." July 23, 2018. https://www.eia.gov/todayinenergy/detail.php?id=36692.

U.S. Environmental Protection Agency. "About Energy Star." https://www.energystar.gov/about.

U.S. Environmental Protection Agency. "Community Connection: Cherry Blossom Bloom Dates in Washington, D.C." Climate Change Indicators. https://www.epa.gov/climate-indicators/cherry-blossoms.

U.S. Environmental Protection Agency. "Overview of Greenhouse Gases." http://www.epa.gov/climatechange/ghgemissions/gases/co2.html.

U.S. Forest Service. *The Rising Cost of Wildfire Operations*. Washington, DC: U.S. Department of Agriculture, 2015.

U.S. Global Change Research Program. *The Climate Report: The National Climate Assessment—Impacts, Risks, and Adaptation in the United States*. Brooklyn: Melville House, 2019.

U.S. Global Change Research Program. *The National Climate Assessment.* Washington, DC, May 2014. http://nca2014.globalchange.gov/report.

U.S. House of Representatives. 116th Congress. 1st Sess. H.R. 109, "Green New Deal." February 7, 2019. https://www.congress.gov/116/bills/hres109/BILLS-116hres109ih.pdf.

U.S. National Hurricane Center. "Saffir-Simpson Hurricane Wind Scale." National Oceanic and Atmospheric Administration. https://www.nhc.noaa.gov/aboutsshws.php.

U.S. National Park Service. "Glaciers—Glacier National Park." October 24, 2013. http://www.nps.gov/glac/forteachers/glaciers.htm.

U.S. National Park Service. "Schoodic Peninsula." https://www.nps.gov/acad/planyourvisit/schoodic.htm.

U.S. National Weather Service. "Summary of Natural Hazard Statistics for 2019 in the United States." June 25, 2020. https://www.weather.gov/media/hazstat/sum19.pdf.

Vandenbergh, Michael, and Jonathan Gilligan. *Beyond Politics: The Private Governance Response to Climate Change.* Cambridge: Cambridge University Press, 2017.

Vanderbilt, Tom. *Traffic.* London: Allen Lane, 2008.

Van Noy, Rick. *Sudden Spring: Stories of Adaptation in a Climate-Changed South.* Athens: University of Georgia Press, 2019.

Vasi, Iono Bogdan. *Winds of Change: The Environmental Movement and the Global Development of the Wind Energy Industry.* New York: Oxford University Press, 2011.

Verba, Sidney, and Norman H. Nie. *Participation in America.* New York: Harper & Row, 1972.

Veron, J. E. N. *A Reef in Time: The Great Barrier Reef from Beginning to End.* Cambridge, MA: Belknap Press of Harvard University Press, 2008.

Victor, David. G. *Global Warming Gridlock: Creating More Effective Strategies for Protecting the Planet.* Cambridge: Cambridge University Press, 2011.

Vince, Gaia. *Adventures in the Anthropocene: A Journey to the Heart of the Planet We Made.* London: Chatto & Windus, 2014.

Vince, Gaia. "Sucking CO2 from the Skies with Artificial Trees." *BBC Future*, October 3, 2012. https://www.bbc.com/future/article/20121004-fake-trees-to-clean-the-skies.

Visser, Steve, and John Newsome. "Alaskan Village Votes to Relocate Over Global Warming." CNN, August 18, 2016. https://www.cnn.com/2016/08/18/us/alaskan-town-votes-to-move/index.html.

Wallace-Wells, David. *The Uninhabitable Earth: Life After Warming*. New York: Tim Duggan Books, 2019.

Wamsley, Laurel. "It Was a Balmy 90 Degrees in Anchorage—For the 1st Time on Record." NPR, July 5, 2019. https://www.npr.org/2019/07/05/738905306/it-was-a-balmy-90-degrees-yesterday-in-anchorage-for-the-first-time-on-record.

Wanless, Herald R. "Sea Levels Are Going to Rise by at Least 20ft. We Can Do Something About It." *The Guardian*, April 13, 2021. https://www.theguardian.com/environment/commentisfree/2021/apr/13/sea-level-rise-climate-emergency-harold-wanless.

Wapner, Paul. "Climate Suffering." *Global Environmental Politics* 14, no. 2 (May 2014): 1–6.

Weart, Spencer R. *The Discovery of Global Warming*. Cambridge, MA: Harvard University Press, 2008.

Weaver, Jacqueline. "The Schoodic Story." *Island Journal*, 2019. https://www.islandjournal.com/history/the-schoodic-story/.

Weinberger, Kate R., Daniel Harris, Keith R. Spangler, Antonella Zanobetti, and Gregory Wellenius. "Estimating the Number of Excess Deaths Attributable to Heat in 297 United States Counties." *Environmental Epidemiology* 4, no. 3 (June 2020): 96. https://journals.lww.com/environepidem/Fulltext/2020/06000/Estimating_the_number_of_excess_deaths.1.aspx?context=LatestArticles.

Weise, Elizabeth. "Climate Change Could Melt Decades Worth of Human Poop at Denali National Park in Alaska." *USA Today*, March 31, 2019. https://www.usatoday.com/story/news/nation/2019/03/31/climate-change-could-soon-melt-years-worth-human-poop-alaska-park/3299522002/.

Weise, Elizabeth, and Rick Jervis. "As Climate Threat Looms, Texas Republicans Have a Solution: Giant Wind Farm Everywhere." *USA Today*, October 18, 2019. https://www.usatoday.com/story/news/2019/10/18/texas-wind-energy-so-strong-its-beating-out-coal-power/3865995002/.

Welch, Craig. "Arctic Permafrost Is Thawing Fast. That Affects Us All." *National Geographic*, September 2019. https://www.nationalgeographic

.com/environment/article/arctic-permafrost-is-thawing-it-could-speed-up-climate-change-feature.

Wetts, Rachel. "In Climate News, Statements from Large Businesses and Opponents of Climate Action Receive Heightened Visibility." *Proceedings of the National Academy of Sciences* 117, no. 32 (August 11, 2020): 19054–60. https://doi.org/10.1073/pnas.1921526117.

White, Christopher. *The Last Lobster: Boom or Bust for Maine's Greatest Fishery?* New York: St. Martin's Press, 2018.

White, Jonathan. *Tides: The Science and Spirit of the Ocean*. San Antonio: Trinity University Press, 2017.

White, Lynn, Jr. "The Historical Roots of Our Ecologic Crisis." *Science* 155, no. 3767 (1967): 1203–7.

White House Office of the Press Secretary. "U.S.-China Joint Announcement on Climate Change." Beijing, China, November 12, 2014. https://www.whitehouse.gov/the-press-office/2014/11/11/us-china-joint-announcement-climate-change.

Wichter, Zach. "Too Hot for Takeoff: Air Travel Buffeted by a Capricious Climate." *New York Times*, June 21, 2017.

Winters, Joseph. "Which Countries Are Responsible for All That Ocean Plastic?" *Grist*, November 2, 2020. https://grist.org/climate/ocean-plastic-which-countries-are-responsible.

Wiser, Ryan, and Mark Bolinger. "2018 Wind Technologies Market Report." U.S. Department of Energy. https://www.energy.gov/sites/prod/files/2019/08/f65/2018%20Wind%20Technologies%20Market%20Report%20FINAL.pdf.

Wolinsky-Hanhias, Yael, ed. *Changing Climate Politics: U.S. Policies and Civic Action*. Los Angeles: CQ Press, 2015.

Wynes, Seth, and Kimberly A. Nicholas. "The Climate Mitigation Gap: Education and Government Recommendations Miss the Most Effective Individual Actions." *Environmental Research Letters* 12, no. 7 (July 12, 2017). https://web.archive.org/web/20200829162700/https://iopscience.iop.org/article/10.1088/1748-9326/aa7541.

Yale Program on Climate Change Communication. "Yale Climate Opinion Maps, 2020." September 2, 2020. https://climatecommunication.yale.edu/visualizations-data/ycom-us/.

Yanarella, Ernest J., and Richard S. Levine. "Research and Solutions: Don't Pick the Low-Hanging Fruit! Counterintuitive Policy Advice for

Achieving Sustainability." *Sustainability: The Journal of Record* 1, no. 4 (August 2008). https://doi.org/10.1089/SUS.2008.9945.

Yoder, Kate. "Why COVID Deniers and Climate Skeptics Paint Scientists as Alarmist." *Grist*, August 13, 2020. https://grist.org/politics/dont-like-what-scientists-are-saying-try-insulting-them.

INDEX